I dedicate this book to my wife Cheree

and to all my grandchildren:

Becky, Jaqueline, Areeg, Hanan, Jarod, Julia, Katheryne, Gabriel, Elena, and Alexander.

Field and Laboratory Techniques in Ecology and Natural History

This Field Techniques Manual is Recommended for High School and College Teachers, and Students

Published by:

North East Naturalist Services
Prospect, CT. 06712
Printed by Book Baby
Author: Alberto F. Mimo MS in Aquatic Biology
With contributions from Anna M. Jalowska Ph.D.
Edited by Jennifer Aley MES
Nenaturalist@sbcglobal.net
www.naturalhistorynotes.blogspot.com
www.northeastnaturalist.com
Library of Congress Registration Number
2018910774
ISBN 978-0-9741411-2-1

If you want to be a biologist
you need to go out of the library.

Table of Contents

What Are Field and Laboratory Technique Manuals?

This field and laboratory techniques manual will provide you, as a teacher, with the opportunity to engage your students in doing a research project.

In the last ten years, science education has been changing from asking students to memorize texts and facts to empowering students to do hands-on research.

It is clear that students should not only memorize facts, but also should be able to process these facts and build on them. Experimentation based on known facts with the objective of learning new things by trial and error is what science is all about.

Over time, we have learned that the *scientific method* is not covered properly in all schools. Many students do not know what the *scientific method* is. And if they do know, they are unable to apply it to real-life scientific projects.

We have also learned that in many cases, teachers are not able to come up with good experiments, and when they do, the methods used in the experiments to complete the research may not be sound and may lack scientific validity.

I have developed a number of field and laboratory techniques throughout my career to provide teachers with the necessary tools to get their students involved in projects that require a hands-on approach and application of the *scientific method.*

I have listed a number of field and laboratory technique applications here ranging from themes in mathematics all the way to techniques in forestry. All the activities are related to ecology and the environmental sciences. Each booklet found on the CD provides you with one application. In each booklet, all found in my website. I give you the information you will need to engage your students in a research project.

I have always said that "*the questions are more important than the answers*". This field and laboratory techniques manual will provide you with a great opportunity to ask good questions and have the students come up with answers without looking them up in a single textbook. The manual will provide you with an introduction, the methods and materials you will need to obtain the results, blank forms to collect the data, and suggestions on how to analyze the data and come up with the results. *But*, let your students analyze the methods and contribute their own grain of sand to the project by finding constructive approaches to improve the methodologies.

Most of these field and laboratory techniques will get the students very involved and should be implemented with plenty of time to let the students think and dissect each project. The results are not as important as the methods used to design the experiments, and the ability of the students to improve the methods. These projects should be done by groups of individuals, and not by one student. Students should be able to discuss the techniques, design their own forms, redesign methods, and have one hundred percent input on the scientific process used to study each case. Let the students organize and direct the outcome of the project. We need to nurture their creativity and allow them to make mistakes. Step back, and let them do the work!

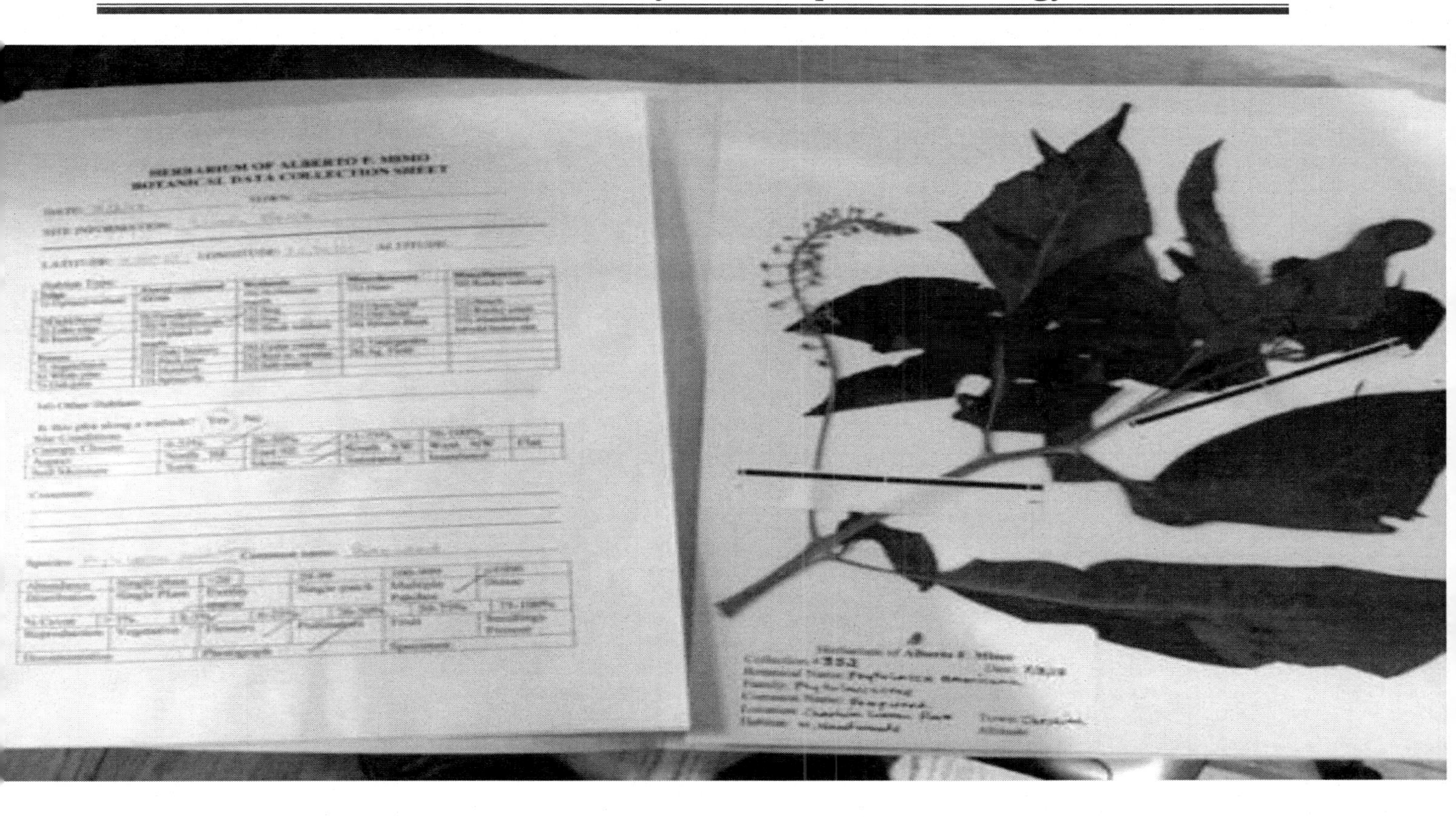

Chapter 1

Name: Working with Specimens and Collecting Data Using a Personal Journal

Activity: Convert your classroom into a museum and teach your students to keep a natural history and laboratory journal.

Level: 1

Developer: Alberto F. Mimo

Site: Outdoors and in the Classroom

Introduction to Collecting

Your classroom should be a natural history museum. Years of working with students and going out into the field should result in a collection of many different and interesting things. But the truth is that these things will be interesting only if you have cared for them, cataloged them and kept them in a safe place, with the help of your students.

I do not have a classroom but my office is a small museum, containing a pretty large collection of fossils. My collection of fresh water crustaceans is so big that it is located in the Collections Department of the University of Connecticut at the Ecology and Evolutionary Biology Department. I have, through the years, collected more than 450 plant specimens that are all mounted and cataloged in an herbarium collection. I also have a large collection of Periphyton samples collected in many streams throughout the Western part of the state of Connecticut with more than 100,000 specimens and more than 200 species. In addition, I have collected insects, copepods, daphnia and much more. Every time I do some research I keep the specimens, do not throw anything away, and catalog and archive everything. Because I work with invertebrates, the only way I can identify them to species is by killing them. I never collect endangered, threatened or species of special concern, and I have state permits to collect.

Education Collecting Permits can be obtained through the Department of Environmental Protection or Fish and Wildlife Agencies.

In field work, catching specimens does not mean we need to keep them. When I do fish inventories I do not keep any. I measure them, take a scale, and send them back on their happy way to the river, lake or ocean. But I keep the data! When I work with Japanese crabs I also collect them, measure them, see where they were living and then put them back in the same location I gathered them from. But keep the data!

Keeping the Catch

Collecting for research purposes can involve many different approaches. If you are doing an inventory and you need to prove that a certain species came from a particular locality, then you need to have documentation. You may need to collect the species and put a sample in a museum. You can take a picture if it was a mammal or a bird but an invertebrate is different, they are too small and you will probably need a microscope to identify them. The same happens with most flowering plants such as goldenrods or asters. You may have the correct genus, but there may be many other species that look alike, so it is important to provide a sample.

Collecting has always been the way that we have inspired kids to become a naturalist. Collecting insects and invertebrates on the ocean beach is a way to get students involved in doing some biology. But on many occasions I see the teachers asking the students to bring 10 plants to the classroom and then after they look at them the whole catch ends up in the waste basket. What a waste!

How about asking each kid to bring one carefully collected plant? Take the plant and do some drawings and pictures, identify the plant, add the plant with all the data, such as location and date collected, into a data base in Excel®, and then add the plant to a class herbarium collection for future reference.

So let's start learning how to do all this the right way!

Field and Laboratory Techniques for Collecting Specimens

Fossil Collecting

I like to collect fossils in the field but that limits my collection because I like to have fossils from all over the world and I will not be able to travel to all those locations. So here is my story about one particular fossil.

1. Most of my fossils come from a rock, mineral and fossil show. The following fossil was acquired in Springfield MA at the Big E show. The show happens every year!

Fossil and Mineral Show in Springfield MA.

2. The fossil was already identified as Weevil Cocoon, or Leptopius duponti, and came with the following information:

3. The next step was to collect as much information as I could find and write it in my fossil journal

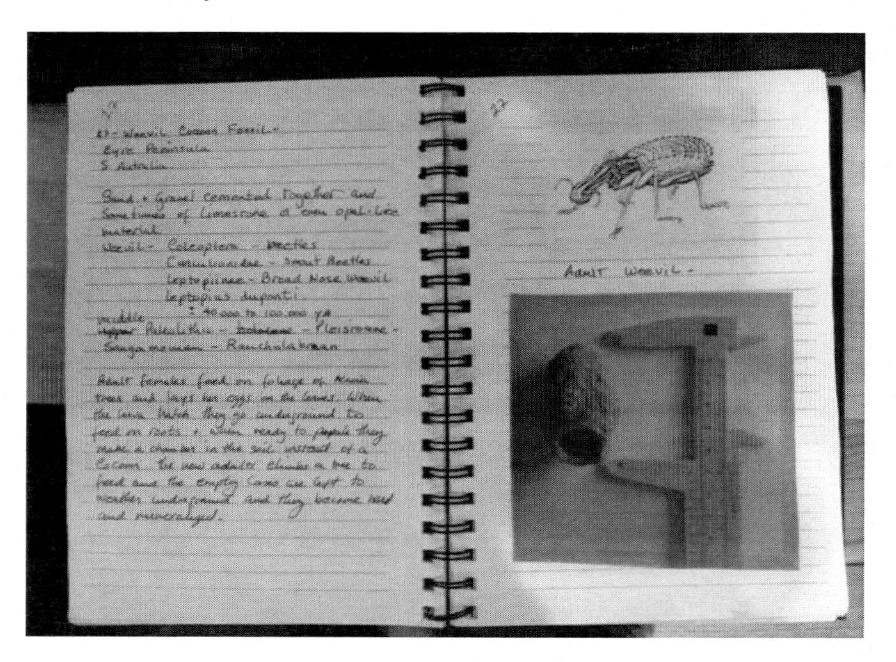

4. When working with students, they need to do all of this. This will teach them to be neat, organized, understand science and how it works and, in the future when they are older, they can go back to your classroom and find their old contributions to your classroom museum. Also by doing the research, they will learn about the organism in natural history including evolution, classification, and so forth. This activity includes writing, reading and art.

Working with Plants

1. Ask your students to collect one plant. Not one leaf, not one stem, but one whole plant and, if possible, with some of the roots. An exception is trees or any shrub.
2. The same day the plants come in to your classroom, they need to be identified. It is much more difficult to do that once they start wilting. Put the plants in some water. There are a number of guides that will help you to identify plants. Here are some examples:

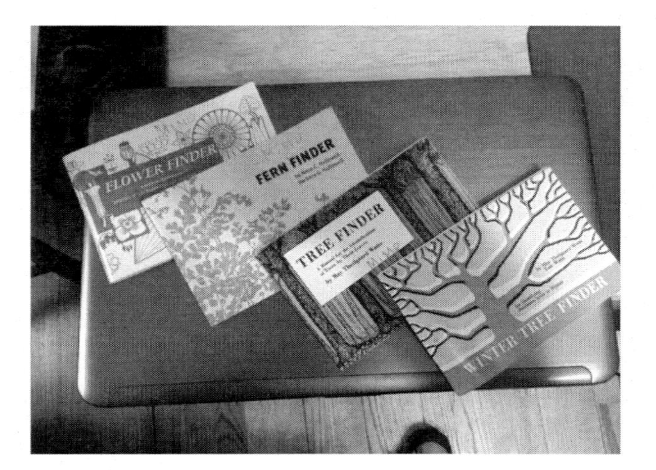

Examples of tree, shrubs, fern and wildflower identification books

3. Once the plant has been identified, the students need to fill out an herbarium form: An herbarium form is at the end of the lesson.
4. You also need to start an Excel® file that will help you keep a data base of the plants you have.

For Example:

Scientific Name	Family	Common Name	Town
Oxalis stricta	Oxalidaceae	Yellow wood sorrel	Torrington
Thlaspi arenses	Brassicaceae	Field pennycress	Derby
Glechoma hideraceae	Lamiaceae	Gill over the ground	Sharon
Prunus spp.	Rosaceae	Ornamental cherry	Danbury
Viola papilionacea	Violaceae	Common blue violet	Danbury
Melilotus officinalis	Fabaceae	Yellow sweet clover	Branford
Solanum dulcamara	Solanaceae	Bitter sweet niteshade	Branford
Rosa multiflora	Rosaceae	Multiflora rose	Branford
Lotus carniculatus	Fabaceae	Birdfoot trefloil	Branford
Petentia erecta	Rosaceae	No common name	Branford

5. Now you need to place your plant on a plant press to dry. Drying will take several months so your plants will be ready to be mounted in the winter.

Plant Press

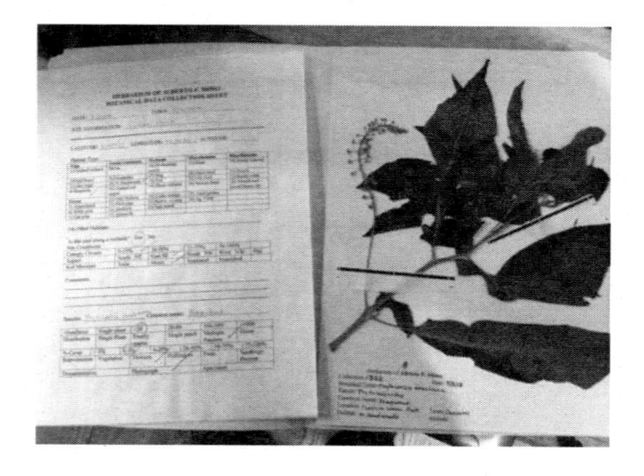

Herbarium form and mounted plant

Mounted plant with the label

6. When a student works with only one plant and learns to identify, label it and look at its natural history, the whole education experience goes much farther. It is not a question of each student bringing many plants; your 30 kids together will create the quantity.
7. In addition to having an herbarium, I also have a journal where I write about each plant and include drawings and photos. Here is an example of an entry in my journal.

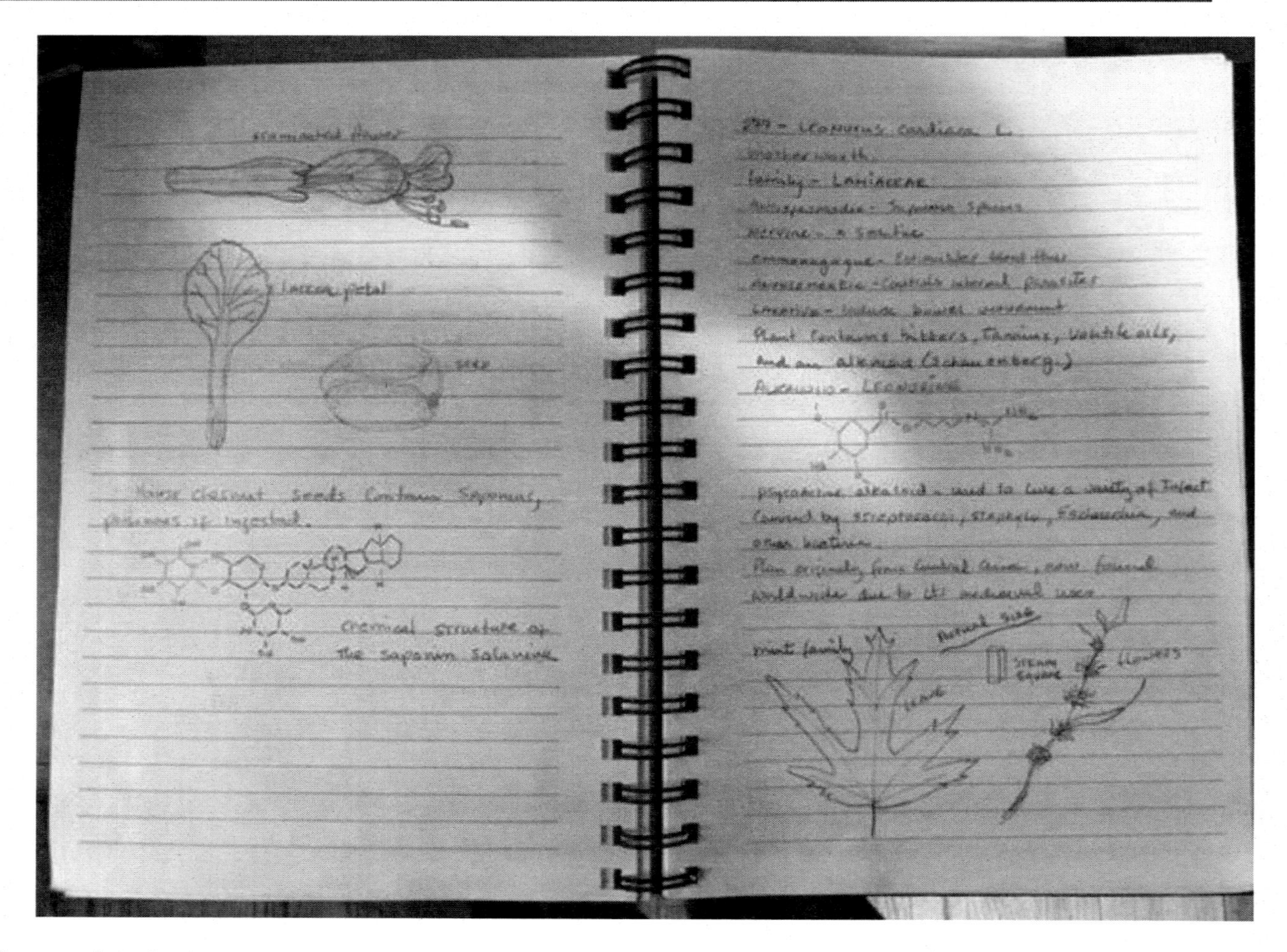

Journal entry

Terrestrial Insects

I use sweep nets to collect insects. Air nets are only good for butterflies and some other delicate insects. Our class objective would be to cover several insect orders. Each order needs to be pinned in a special way. Pinning insects is an art. There are several web sites with instructions on how to do this.

https://www.homesciencetools.com/content/reference/IN-INSEPIN.pdf

https://www.youtube.com/watch?v=MT5VGISCtg4

http://extension.oregonstate.edu/umatilla/sites/default/files/PINNING__INSECTS.pdf

https://bugguide.net/node/view/36900

Identifying insects to the order is easy. I recommend using the "Peterson Guide to Insects" or any other insect guide. There are field guides to the butterflies, to the moths, to the ants, to the dragonflies and much more.

BOTANICAL DATA COLLECTION SHEET

DATE:______________________ TOWN: ____________________________

SITE INFORMATION: ____

LATITUDE: __________ LONGITUDE: ___________ ALTITUDE: __________

Habitat Type:

Edge	**Forest continued**	**Wetlands**	**Miscellaneous**	**Miscellaneous**
1) Upland/wetland	8) Oak	16) Herbaceous marsh	23) Dune	30) Rocky outcrop
2) Field/forest	9) Floodplain	17) Bog	24) Open field	31) Beach
3) Lake edge	10) Northern Hardwoods	18) Fen	25) Old field	32) Rocky coast
4) Roadside	11) Upland red maple	19) Shrub wetland	26) Stream Bank	33) Abandoned lot/old home site
Forest	12) Oak / Hickory	20) Cedar swamp	27) Yard/garden	
5) Aspen/birch	13) Pitch pine	21) Red maple swamp	28) Agricultural Field	
6) White pine	14) Hemlock	22) Salt marsh		
7) Oak/pine	15) Spruce/fir			

34) Other Habitats: ____

Is this plot along a trailside? Yes No

Site Conditions

Canopy Closure	0-25%	26-50%	51-75%	76-100%	
Aspect	North NE	East SE	South SW	West NW	Flat
Soil Moisture	Xeric	Mesic	Saturated	Inundated	

Comments:

____

____

____

Species: ______________________ Common name: ____________________________

Abundance	Single plant	>20	20-99	100-999	<1000
Distribution	Single plant	Evenly sparse	Single patch	Multiple patches	Dense

% Cover	> 1%	1-5%	6-25%	26-50%	50-75%	75-100%

Reproduction	Vegetative	Flowers	Pollinators	Fruit	Seedlings Present

Documentation	Photograph	Specimen

Working with Aquatic Macroinvertebrates and Plankton

The idea is the same. Collect them, keep them in a safe place, identify them, add them to your journal and Excel® data base, but in the case of macroinvertebrates, you need to preserve them in alcohol. I buy 90% rubbing alcohol from the pharmacy; it is cheaper than laboratory grade ethanol and works just as well.

Insects and other macroinvertebrates are very difficult to identify to the ***species*** except for some rare cases. Family or genus level should be sufficient. What is more important is to keep complete communities as they were found together living in the same conditions. In many cases I keep them in the same vial with information about location, the way they were collected and the date. If you go to the same stream every year with different classes, you can study the changes in specimen composition, and learn about the site and what is happening there. It is always important to know: where the specimens come from, some knowledge of the site, when they were collected, and in the case of fossils you also need to know the geologic formation.

Some specimens such as jelly fish need to be anesthetized before preservation, some specimens need to be fixed with 10% formalin and some need to be dried. Look it up on the web for details. Biological companies sell nontoxic preservatives.

Journal Keeping

Yesterday I saw a documentary about Jane Goodall. In that documentary, they show how she took notes on the behavior of chimpanzees and other animals. Taking notes is not easy. It is important before taking notes to make a list of the areas that we want to observe. Observations may have to do with behavior or physical observation of anatomy and physiology, details about the surroundings, weather, relationships between organisms, and a variety of ecological details.

In most cases, students, and also researchers, use data sheets to write the results of their investigations. One paper, one clipboard and a pencil is all you need. When you come back from the field you need to transfer the results to a journal. Pieces of paper get lost, dirty, crumpled and ruined but a journal is always neat and safe.

I do not use data sheets; all observations go into a field journal that I keep forever. I write about the many interesting things that I encounter in the field such as mammal or bird sightings, a weather-related event, difficulties with the collection, and small details that will never be written on a data sheet.

Field Journal

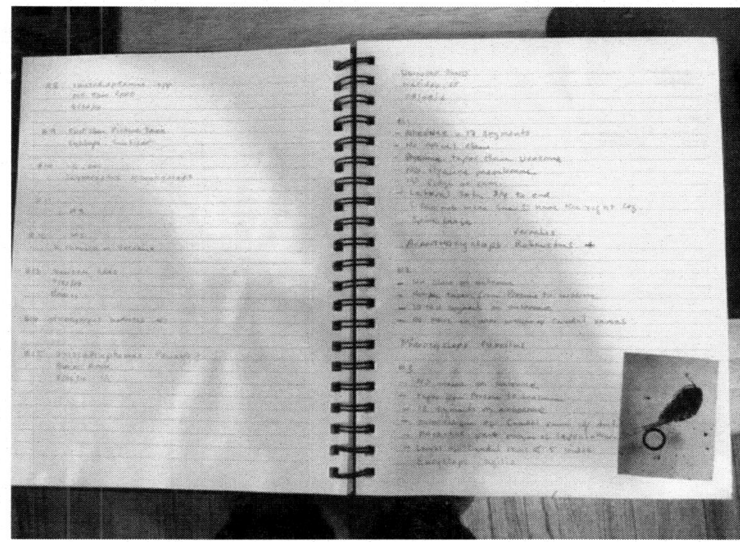

Laboratory Journal

If you have an art inclination I suggest you read “Keeping a Nature Journal” by Clare Walker Leslie and Charles E. Roth. This book will provide you with many ideas. You should also team up with your math and art teacher.

Many times I have done research with high school students and I usually get cheap notebooks for them. Unfortunately, I do not have these students all year, just for a short time, but I always encourage them to write, and draw pictures, and make maps, and use color and graphs.

Activity

The objective of this activity is to start the journals and complete a small collection.

1. For this activity, your students will need to have a journal. They can buy their own or you can purchase journals for all of them. If the students buy their own it is cheaper and their journals will be selected by them and reflect their own personalities.
2. To start their journals ask them to decorate the first page, and write their name and a way to return the journal if it is lost and found by someone else.
3. To start this lesson I suggest going through the basics of leaf morphology in plants. Alternate, opposite, serrated, entire, parallel venation, and much more. Students are to take notes and do drawings in the journals.

4. Divide the students into 4 or 5 groups that will work together, but each student should do their own journal writing using feedback from their group.
5. Go out into the field, around the school and collect 10 plants (trees and shrubs) per group. Be careful not to collect ornamental plants because they are difficult to identify.
6. Bring the plant samples into the classroom and study them. Complete drawings and sketches in the journal, including physical details such as measurements of the leaves, stem, and the whole plant. Are these plants evergreen or deciduous? Are they a tree or a shrub?
7. Using a dichotomous system ask the students to use the physical qualities of each plant, such as the shape of leaves, edges, alternate, parallel or other growth patterns, venation, and color, to make a key. Let the students name the plants using invented names.
8. Using a real plant key identify the plants by name. Compare results.
9. Use the herbarium form and fill out the questions.
10. Press the plants and, later on in the year, mount them on herbarium paper.

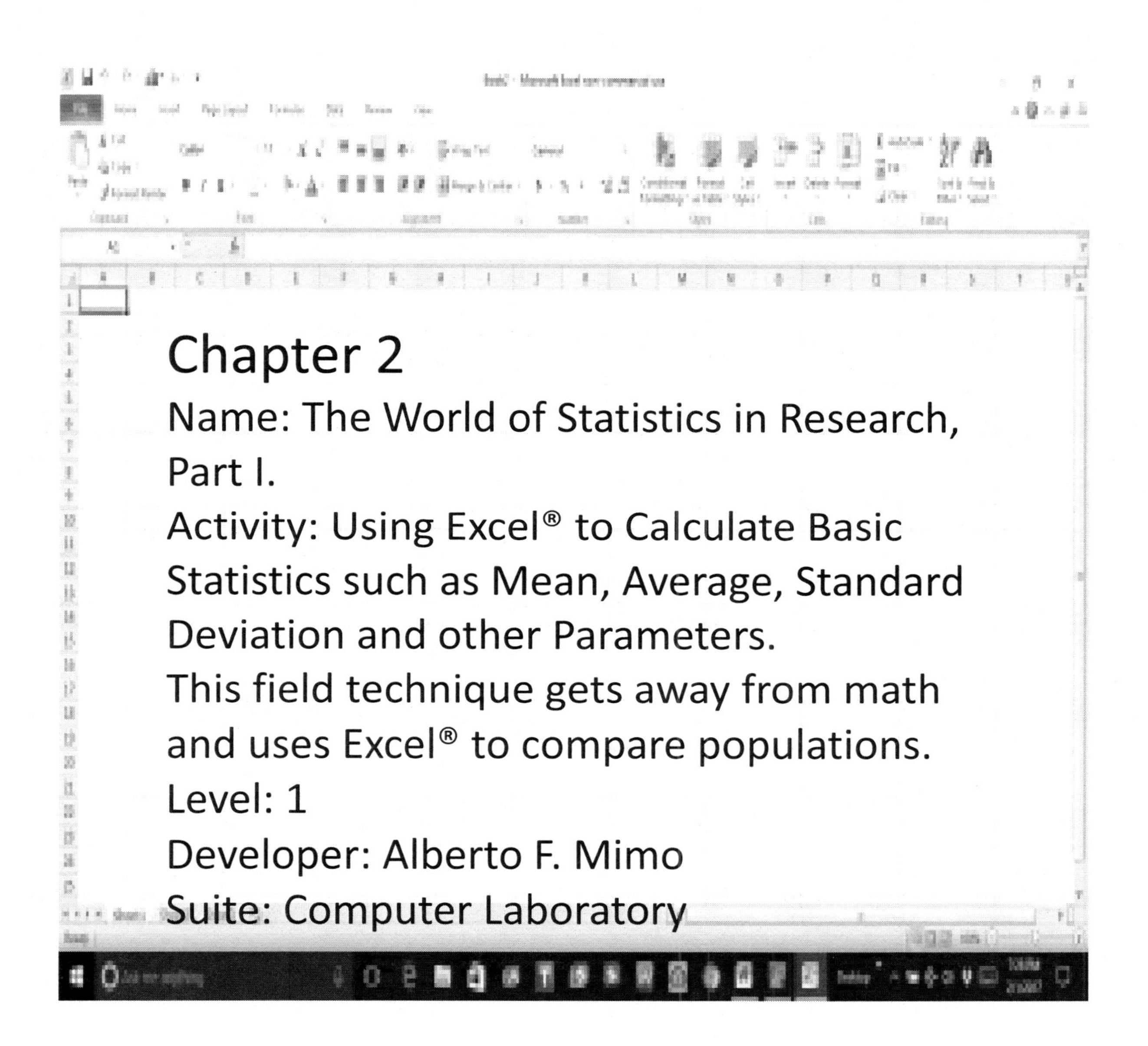
Chapter 2
Name: The World of Statistics in Research, Part I.
Activity: Using Excel® to Calculate Basic Statistics such as Mean, Average, Standard Deviation and other Parameters.
This field technique gets away from math and uses Excel® to compare populations.
Level: 1
Developer: Alberto F. Mimo
Suite: Computer Laboratory

Introduction: What are Statistics For?

“Stats” have been used for years to help scientists evaluate the accuracy of their data, choose between different numbers, improve the certainty that the data is right, learn more about probabilities and much more, but… “stats” have also been used to make scientist appear right when they were not! So, working with "stats" is something that has to be done carefully. You do not want people to say that you misused "stats" to prove your point!

Scientists today often do not work out of a hypothesis. They usually design an experiment, with the intentions of learning more about something. Then they try to explain their results by using natural logic and their collected data. It is true that scientists have a hypothesis but usually, they work on a subject and to learn more they execute an experiment.

Example:

Entomologists are interested in the feeding habits of ants. No scientist will try to predict what ants eat and then check if they do. They will set up an experiment where they will be able to observe the ant’s behavior and after collecting the data many times, for many ants and in many places, they will describe their observations and provide us with an explanation regarding the feeding habits of ants. Scientists are not engaged in a guessing game.

We cannot say the ants eat seeds because we once saw an ant eat a seed. Good science has to be done over a period of time and using repeated measurements and observations.

Don’t Worry, we are not sending you out to figure out what ants eat!

Fun Starts Here!

The World of Statistical Words

It is important that before you use statistics you learn and understand some vocabulary words.

Population - A population is considered to be a number of individuals all from the same species.

Community - In ecology, the community is the overall set of organisms and abiotic conditions existing within one habitat together. Your community includes weather conditions, soil chemistry, sunlight at a particular site and so on. All plants and animals are also part of a community.

Sample size - The number of observations, which include missing and non-missing values. Zeros are also counted.

Missing - The number of missing observation values.

Maximum - The maximum value of the sample observations.

Minimum - The minimum value of the sample observations.

Sum - The sum of the values of all the sample observations.

Mean - The arithmetic average of the sample values. It is the sum of the values divided by the sample size.

Median - The middle value (or arithmetic mean of two middle values) of the sample values when they are sorted in order. For sorted data, if "n" is odd, then the median is equal to the center value. If "n" is even, then the median is equal to the average of the two center values.

Variance - This measures the dispersion of a sample. It is one of the most widely accepted measures of the variability of a set of data.

Standard Deviation - Another popular measurement of the dispersion around the mean. It is defined as the square root of the variance.

Confidence level- The maximum and minimum number expected for that population.

It is my advice that you stay away from making students complete cumbersome mathematics. Use an Excel® program and teach the kids how to use it. The objective here is not to be able to do the math. The objective here is to use STATISTICS to better understand your scientific results. Graph, Graph, Graph!

Using Excel®

1. This is what an Excel® sheet looks like.

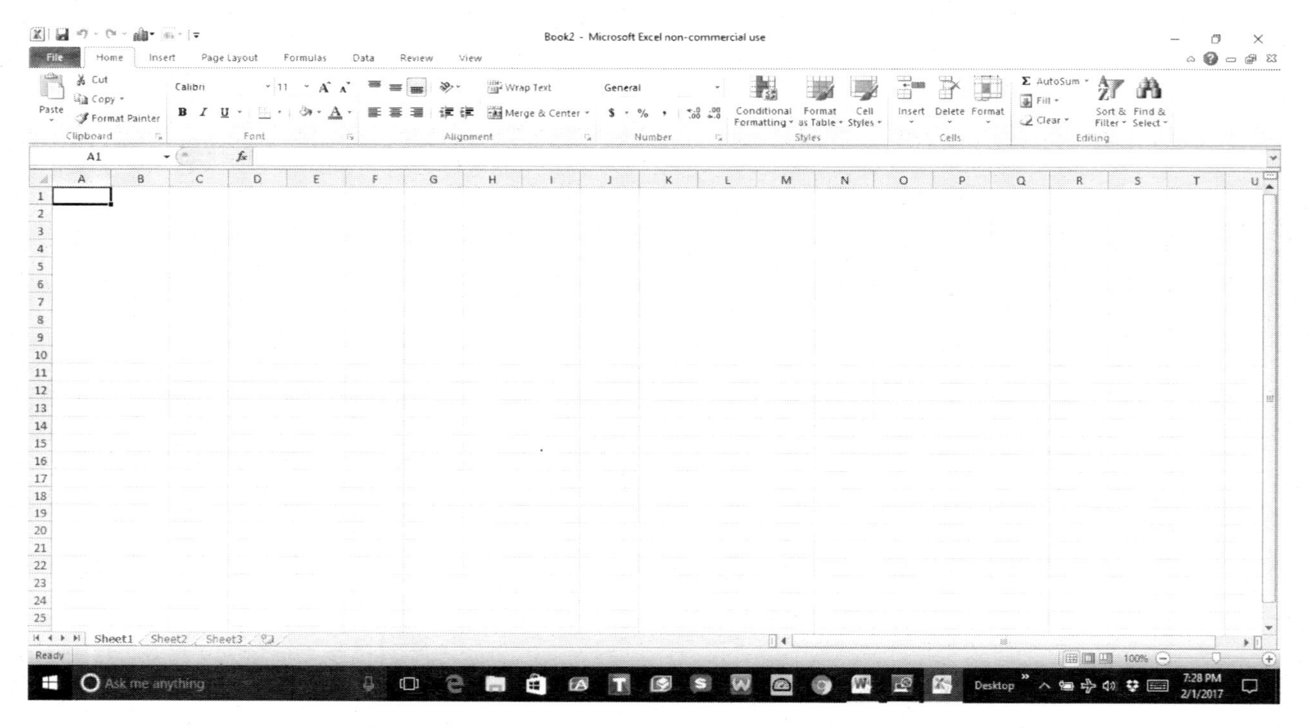

All examples use Excell®

2. You can enter numbers in each row, so if I have a bunch of numbers I can place them in that row.

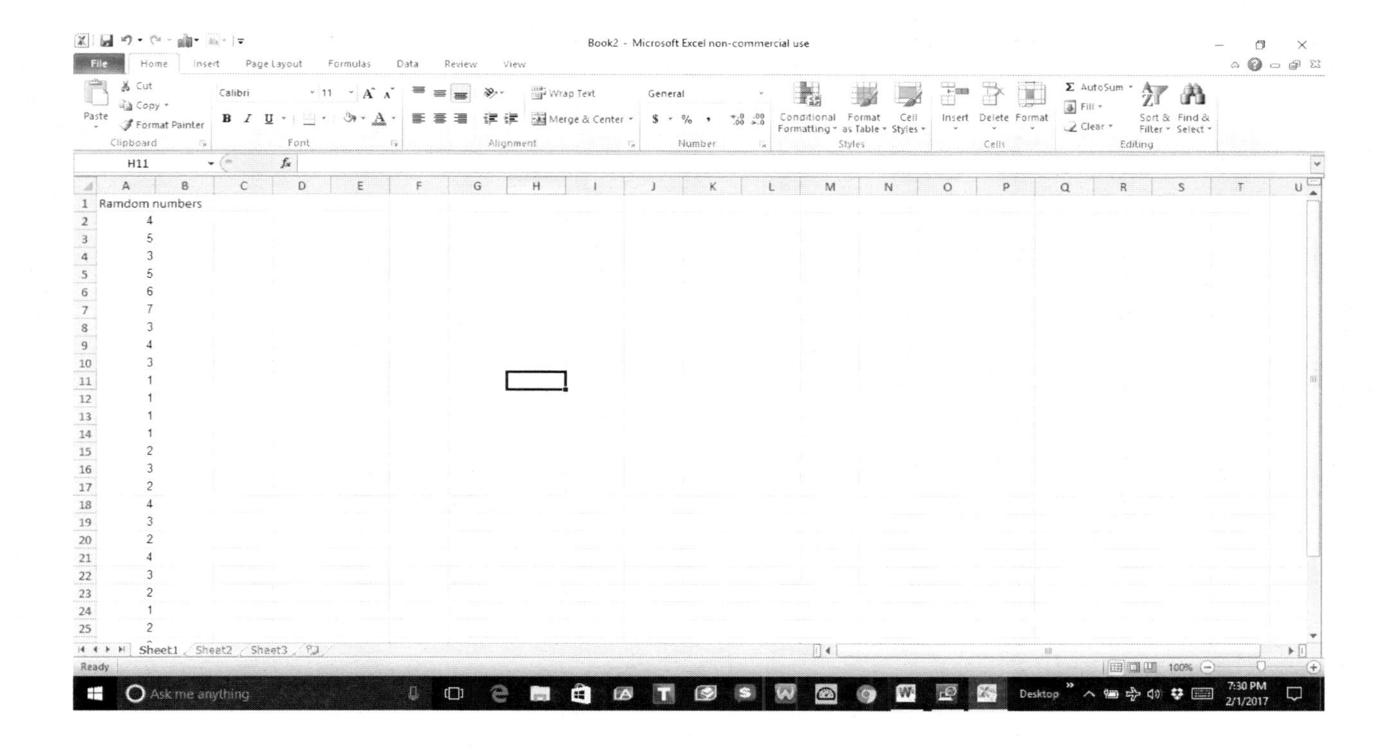

3. Now I can get the program to do some calculations for me. First, we need to add in stats to your excel program. So go in the program File and click options.

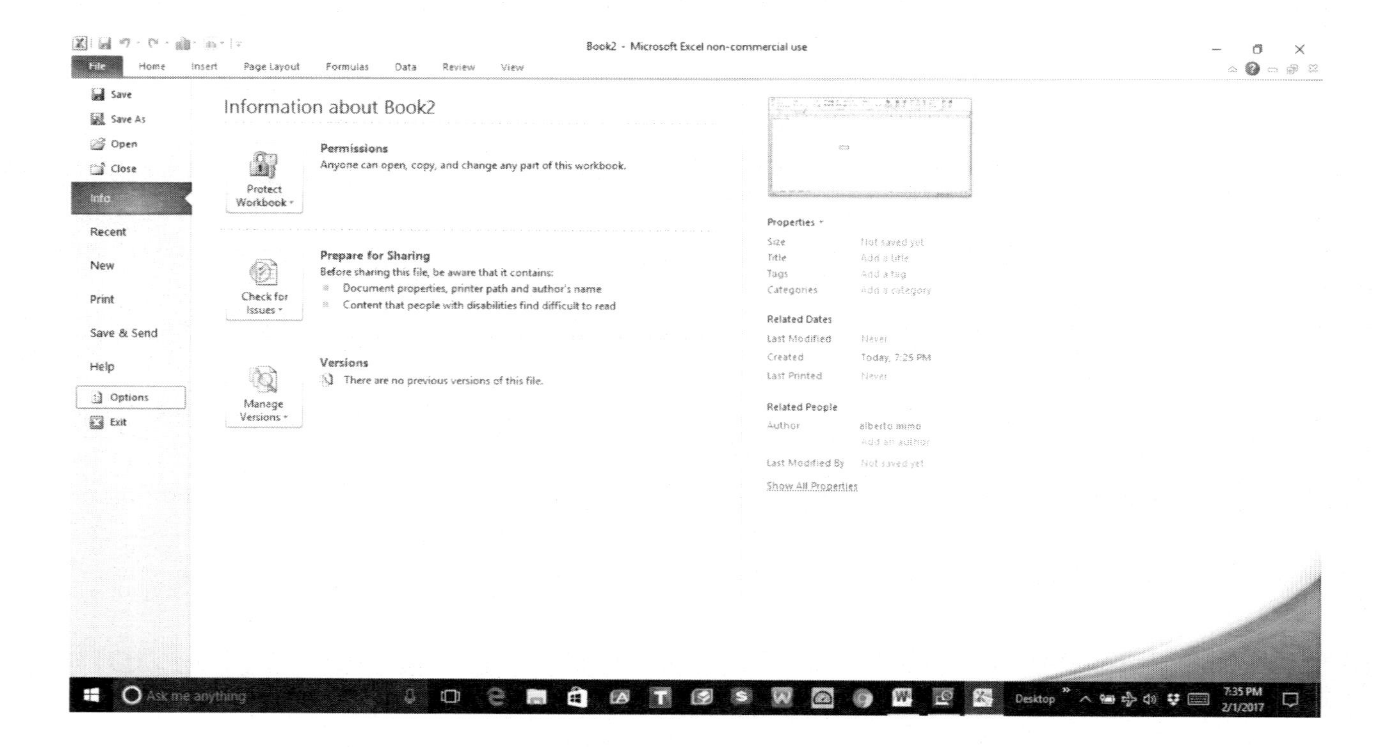

4. To add statistical calculations click on add-ins.

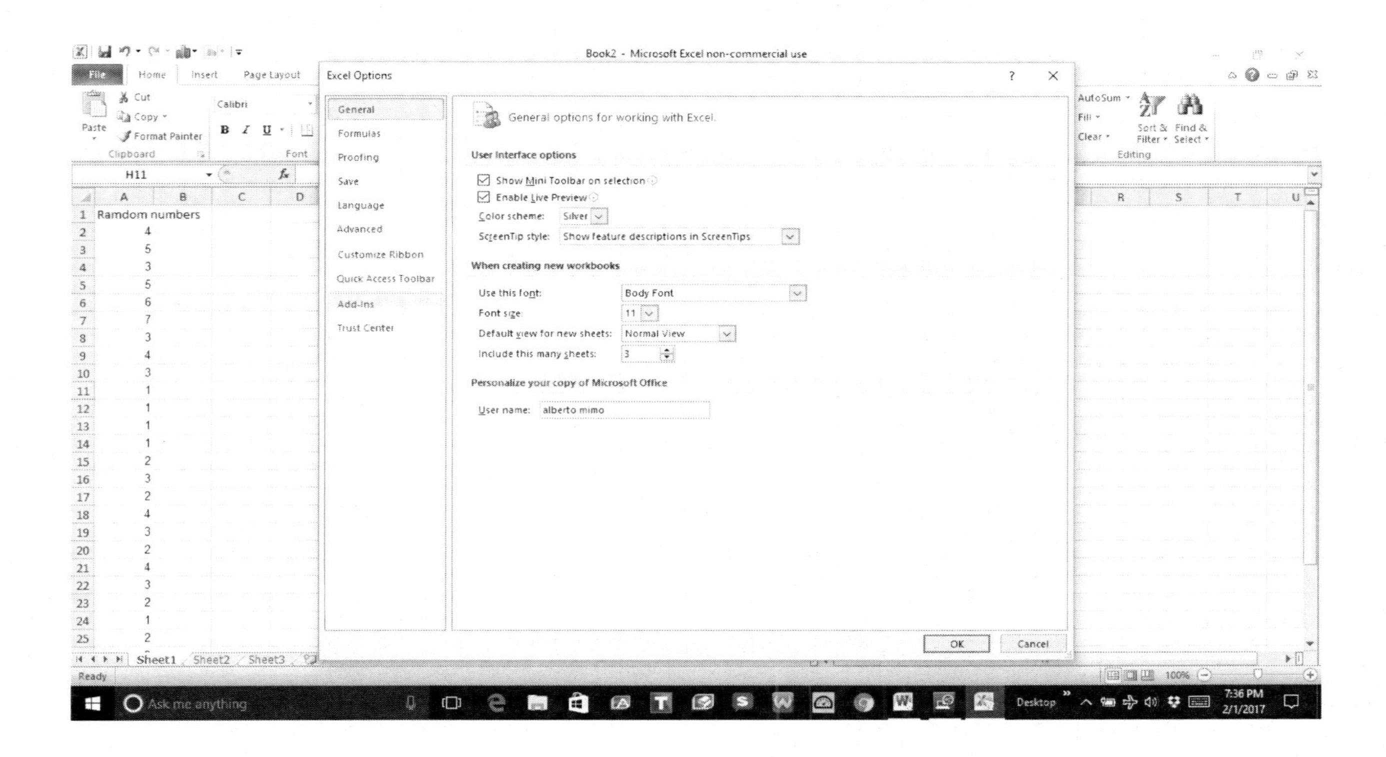

5. Once you have the next screen add the analysis tool pack.

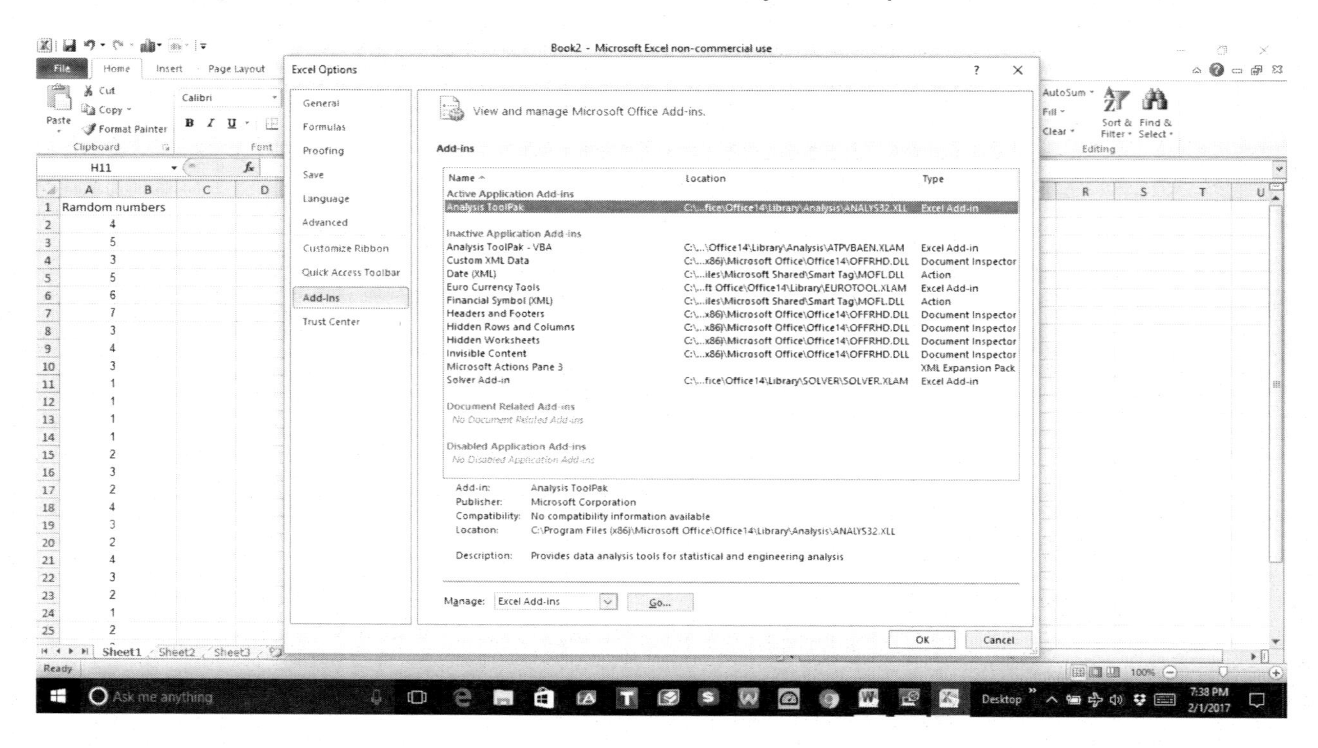

6. Click OK and when you go back to the main page and click on Data, you will see the addition on the right hand side corner.

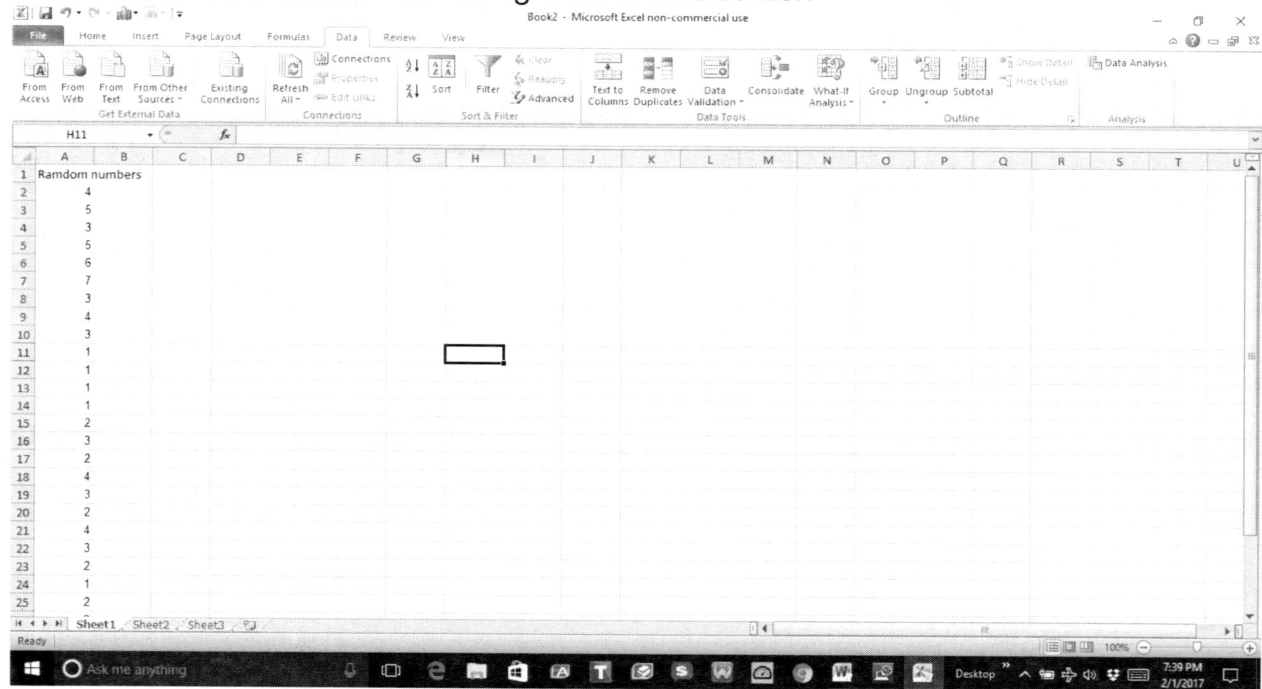

7. This feature is easy to use. We will do it together.

Let's do the MATH

1. The first thing we'll do is to learn how the program works by doing some math with a random set of numbers just to learn how to use excel. Our next step is to ask excel to do some basic stats, and we'll work from that point to understand what each statistic number means. So let's enter into one of the columns 20 random numbers between 1 and 100.

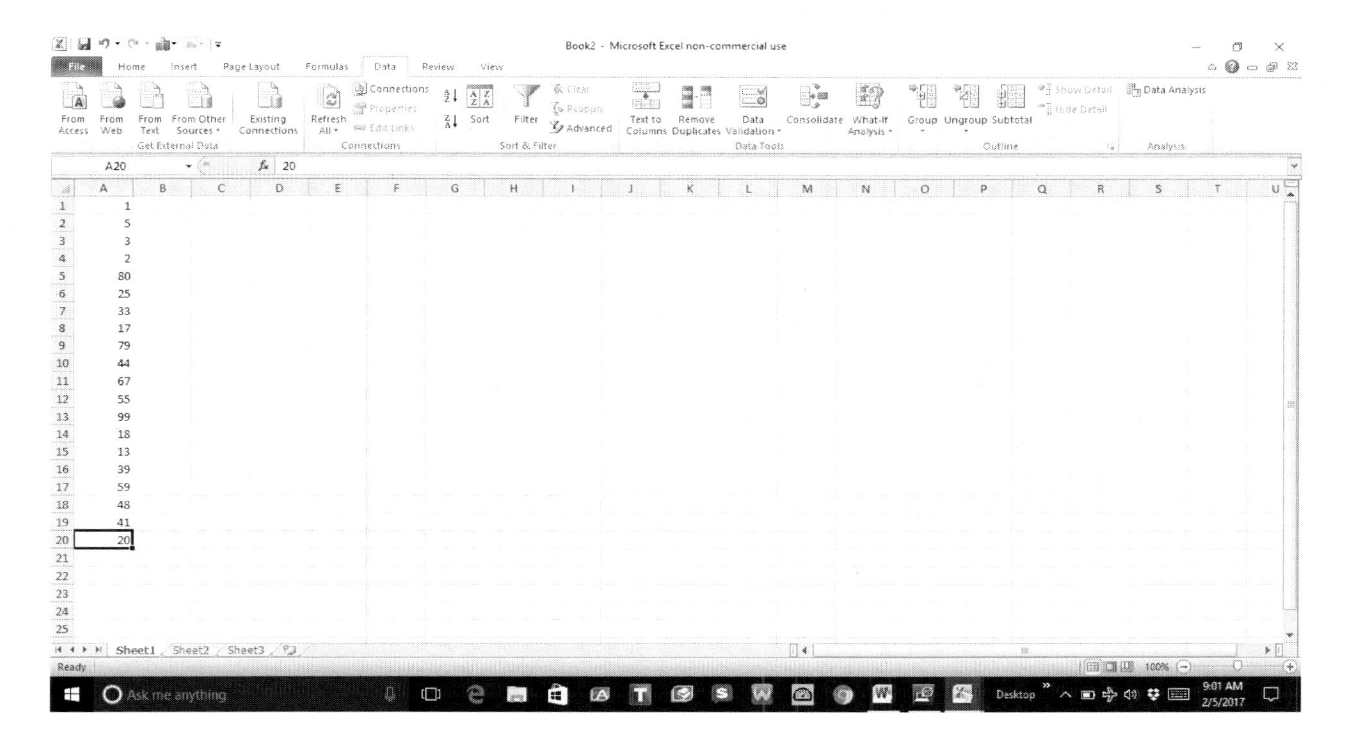

2. To complete a stat analysis we will click on "Data Analysis" and follow the directions.

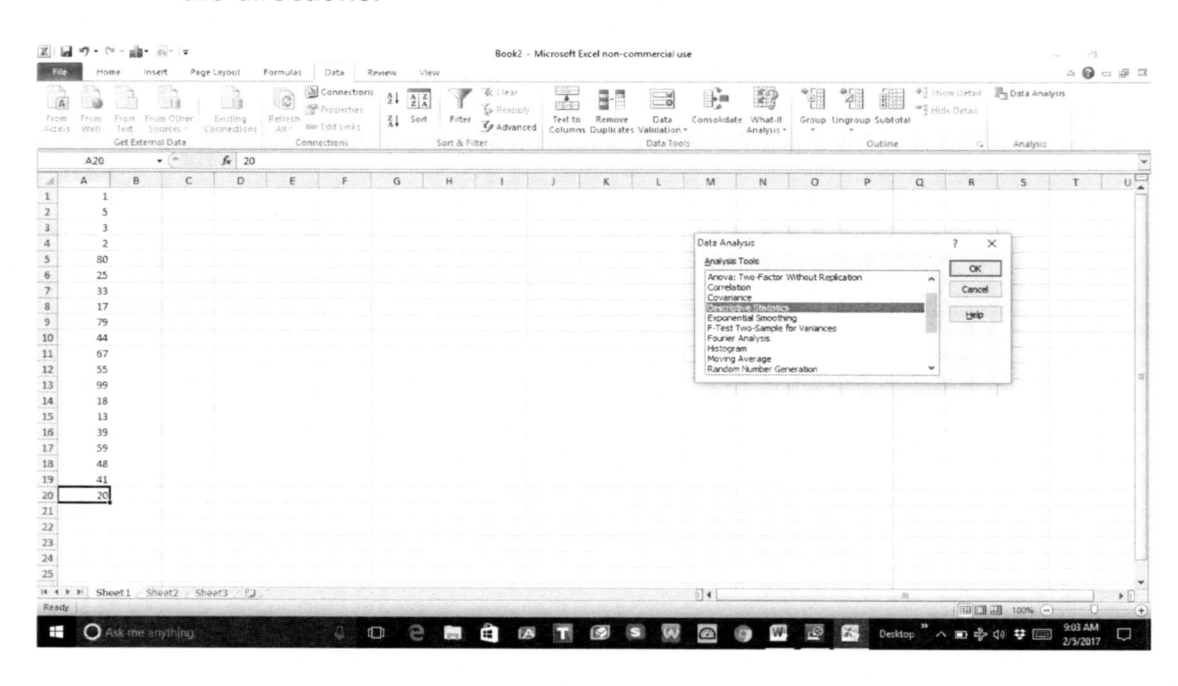

3. Choose "Descriptive Analysis", and click OK. For "input range", we will click on the icon and choose all the numbers in our column.

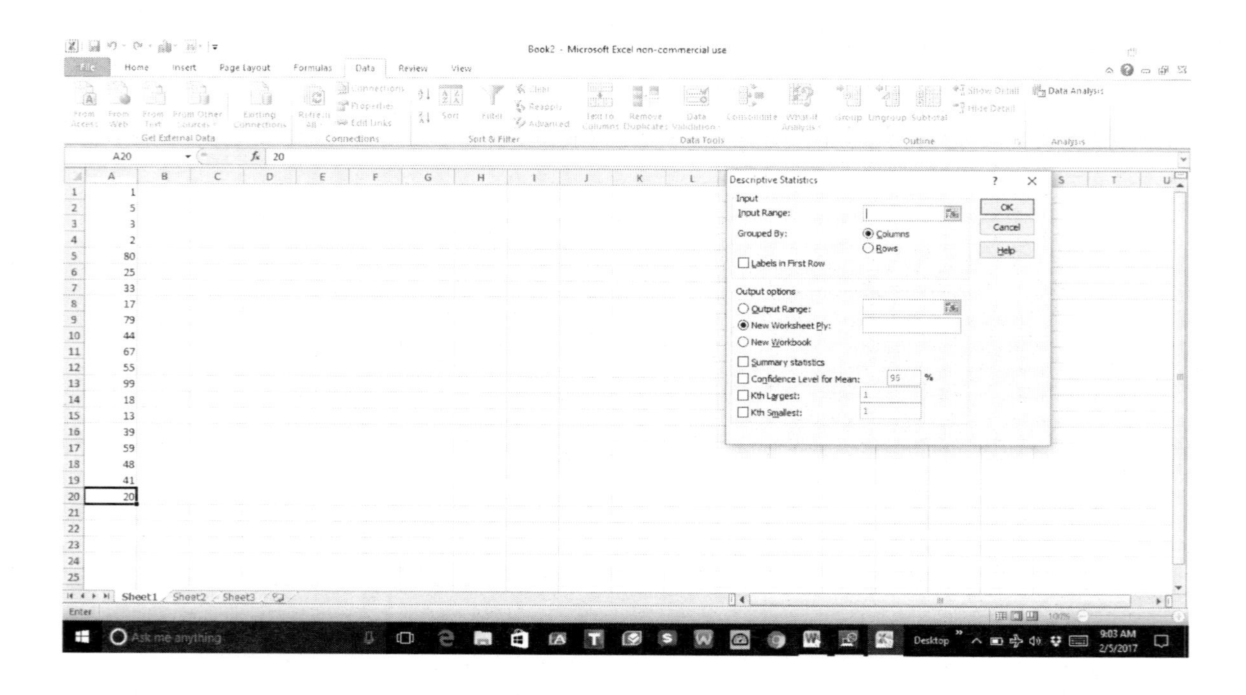

4. Next, we will check the summary statistics and ask Excell® to also give us a 95% confidence level for the mean. Then click OK.

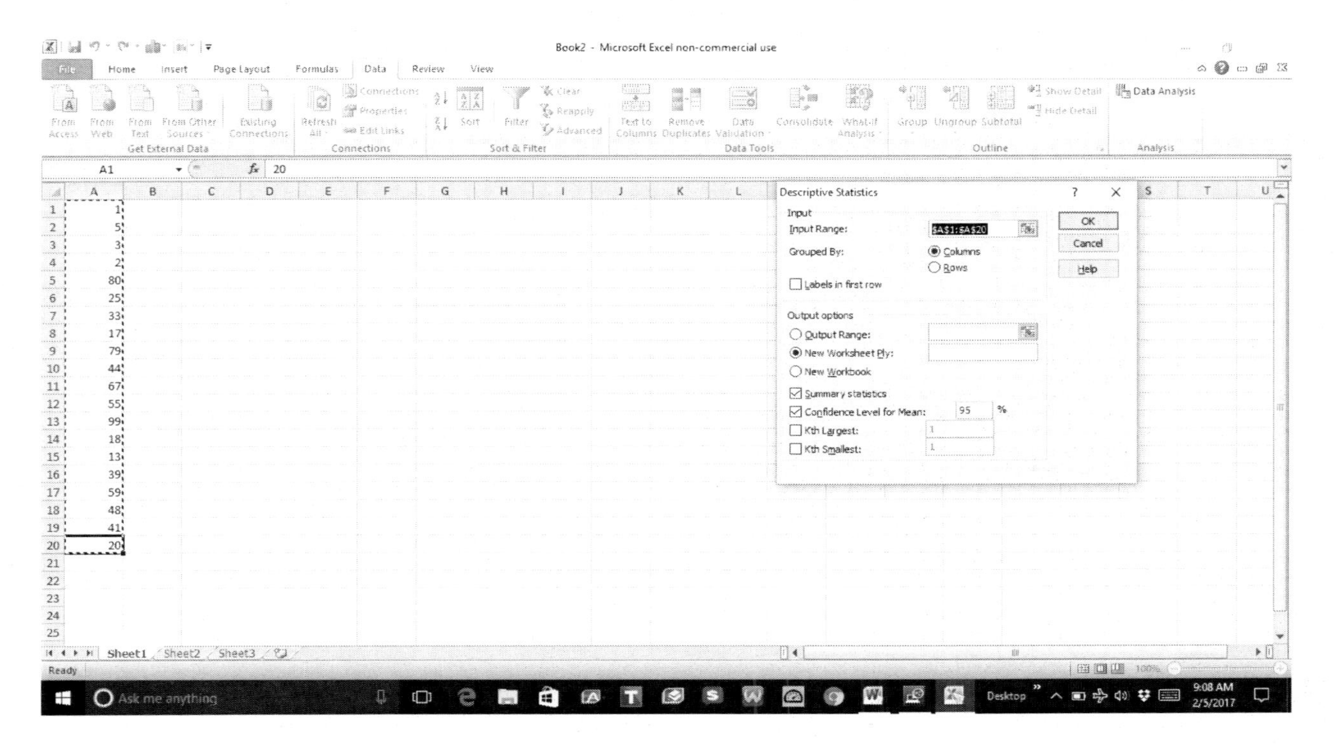

5. On sheet 4 we have all our results.

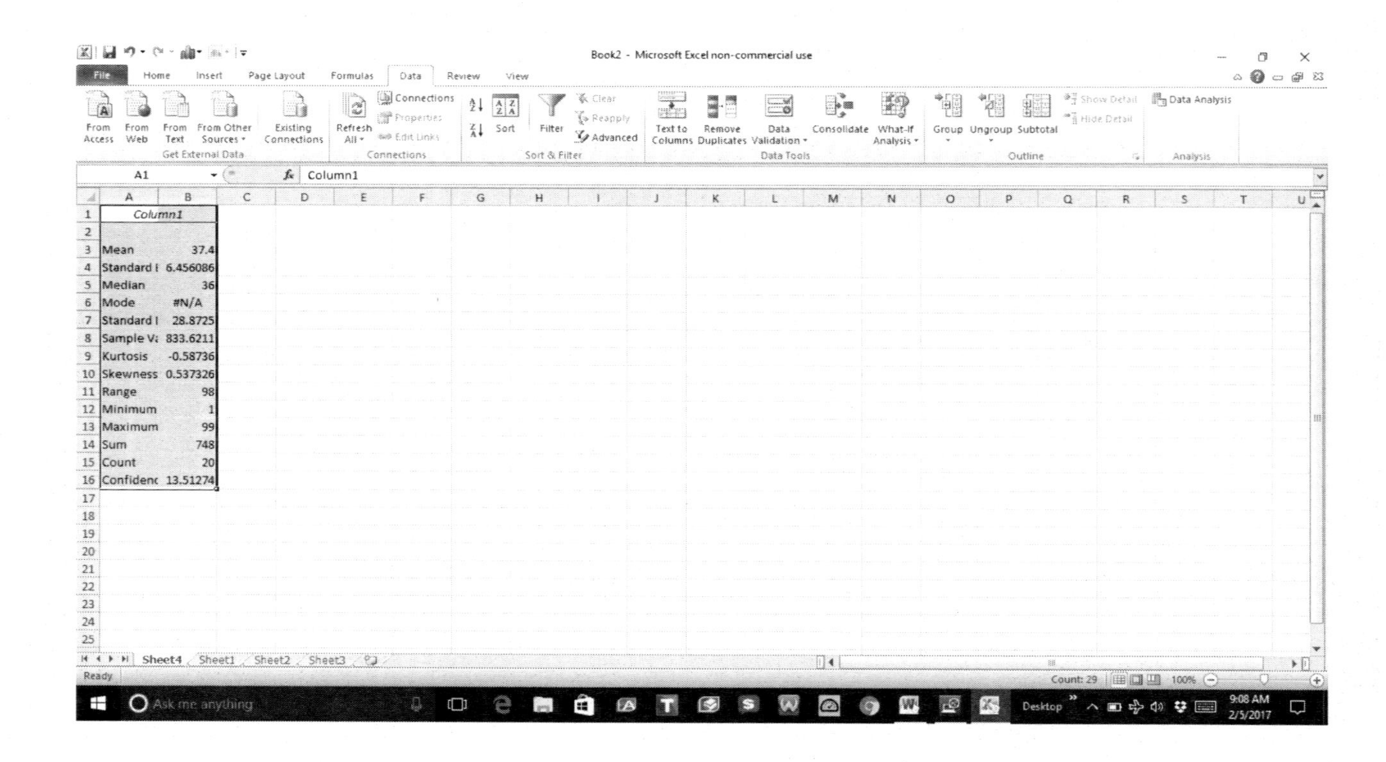

What does it all mean?

Here we have our results without having to do the math. It is important that we concentrate on these numbers and learn what they mean. Let's copy and paste these numbers onto sheet #1.

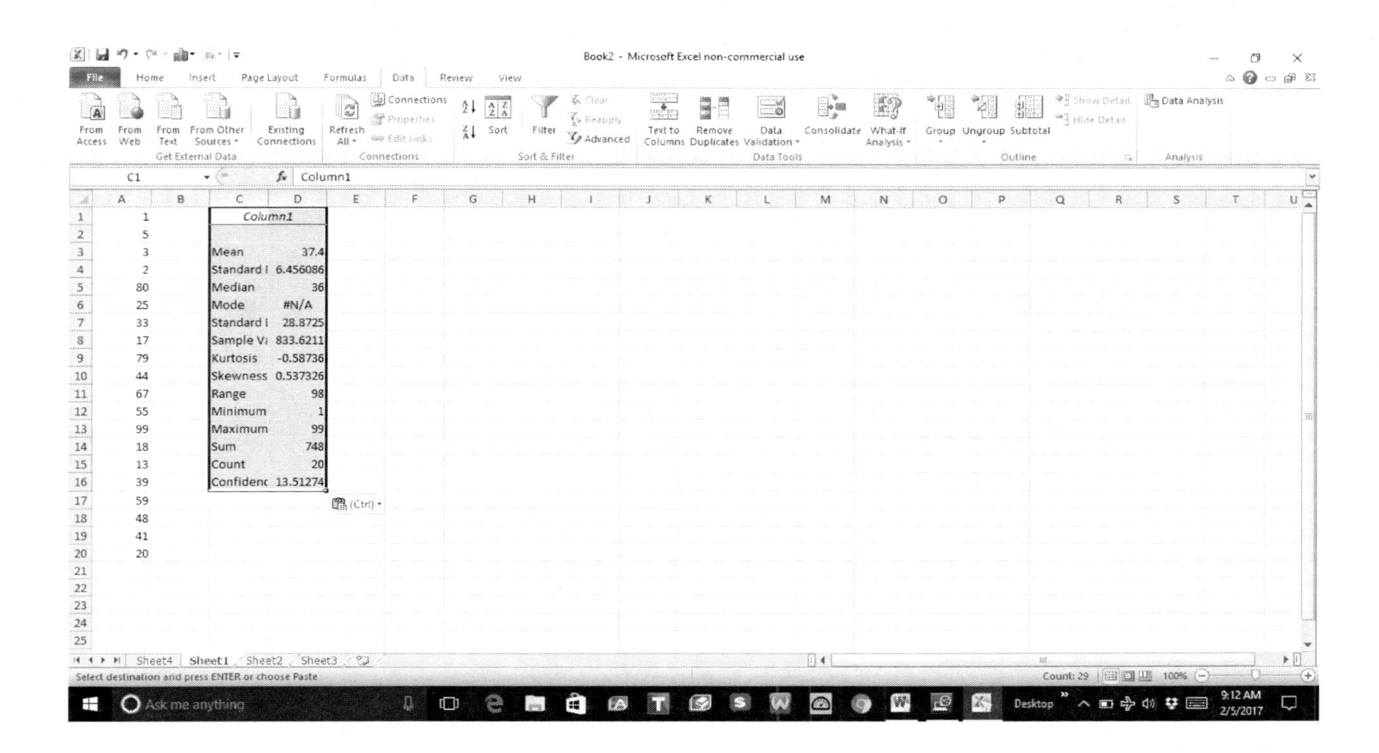

	A	B	C	D
1	1		*Column1*	
2	5			
3	3		Mean	37.4
4	2		Standard I	6.456086
5	80		Median	36
6	25		Mode	#N/A
7	33		Standard I	28.8725
8	17		Sample Vi	833.6211
9	79		Kurtosis	-0.58736
10	44		Skewness	0.537326
11	67		Range	98
12	55		Minimum	1
13	99		Maximum	99
14	18		Sum	748
15	13		Count	20
16	39		Confidenc	13.51274
17	59			
18	48			
19	41			
20	20			

Usually to work with and understand statistics it is best to use a real experiment. Learn how to complete an experiment and use your statistics in the next Field and Laboratory Techniques Chapter "The World of Statistics".

Chapter 3

Name: The World of Statistics in Research, Part II.

Activity: Using a population ecology research example to work with a crustacean population.

Level: 2

Developer: Alberto F. Mimo

Site: Computer Laboratory

Introduction

In the world of Statistics Part II, I will compare the number of crabs found in a ¼ square meter, using samples of two populations of Japanese Crabs (*Hemigrapsus sanguineous*). The first population of crabs comes from Hammonasset Beach State Park, and the second one is made up of all the crabs found in Connecticut by all the teams that worked on this project. (Project sites included Calf Park in Stamford, Sherwood Island in Westport, and New Haven Harbor, plus Hammonasset Beach State Park in Madison) Project teams sampled the numbers of males and females found in a ¼ square meter at low tide in an area no more than 25 feet from the water's edge.

In the following example I have copied and pasted my raw information from the two different groups into an Excel® sheet and then completed a **"Descriptive Statistics Analysis"** on each set (See The World of Statistics: Part I). I have a total of four sets, the number of Males and Females from Hammonasset Beach State Park and the number of Males and Females from all the sites. (To work with a two row set of numbers just select both rows in Excel®, the program does the rest automatically)

Examples

Here is the magic of statistics, NO MATH needs to be done by you! The Excel® program will do it all for you.

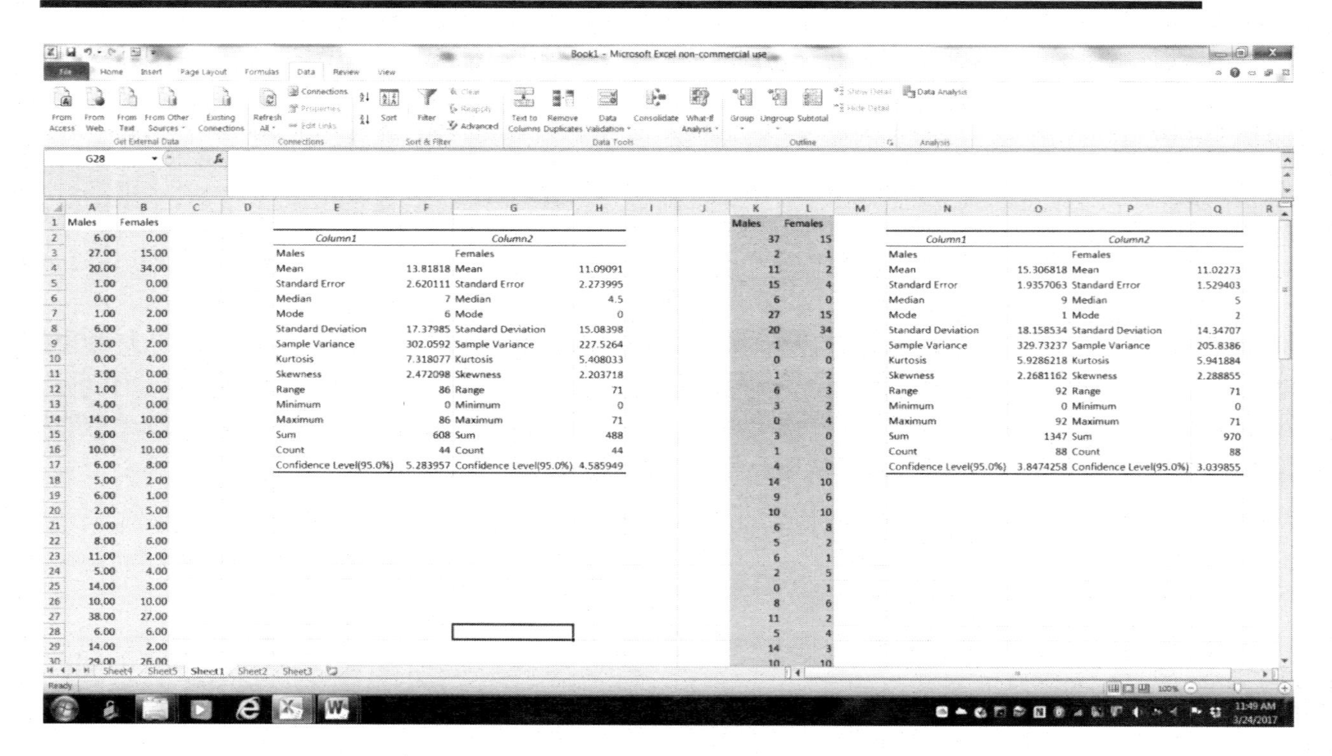

Row	Males	Females
2	6.00	0.00
3	27.00	15.00
4	20.00	34.00
5	1.00	0.00
6	0.00	0.00
7	1.00	2.00
8	6.00	3.00
9	3.00	2.00
10	0.00	4.00
11	3.00	0.00
12	1.00	0.00
13	4.00	0.00
14	14.00	10.00
15	9.00	6.00
16	10.00	10.00
17	6.00	8.00
18	5.00	2.00
19	6.00	1.00
20	2.00	5.00
21	0.00	1.00
22	8.00	6.00
23	11.00	2.00
24	5.00	4.00
25	14.00	3.00
26	10.00	10.00
27	38.00	27.00
28	6.00	6.00
29	14.00	2.00
30	29.00	26.00

Column1		Column2	
Males		Females	
Mean	13.81818	Mean	11.09091
Standard Error	2.620111	Standard Error	2.273995
Median	7	Median	4.5
Mode	6	Mode	0
Standard Deviation	17.37985	Standard Deviation	15.08398
Sample Variance	302.0592	Sample Variance	227.5264
Kurtosis	7.318077	Kurtosis	5.408033
Skewness	2.472098	Skewness	2.203718
Range	86	Range	71
Minimum	0	Minimum	0
Maximum	86	Maximum	71
Sum	608	Sum	488
Count	44	Count	44
Confidence Level(95.0%)	5.283957	Confidence Level(95.0%)	4.585949

Males	Females
37	15
2	1
11	2
15	4
6	0
27	15
20	34
1	0
0	0
1	2
6	3
3	2
0	4
3	0
1	0
4	0
14	10
9	6
10	10
6	8
5	2
6	1
2	5
0	1
8	6
11	2
5	4
14	3
10	10

Column1		Column2	
Males		Females	
Mean	15.306818	Mean	11.02273
Standard Error	1.9357063	Standard Error	1.529403
Median	9	Median	5
Mode	1	Mode	2
Standard Deviation	18.158534	Standard Deviation	14.34707
Sample Variance	329.73237	Sample Variance	205.8386
Kurtosis	5.9286218	Kurtosis	5.941884
Skewness	2.2681162	Skewness	2.288855
Range	92	Range	71
Minimum	0	Minimum	0
Maximum	92	Maximum	71
Sum	1347	Sum	970
Count	88	Count	88
Confidence Level(95.0%)	3.8474258	Confidence Level(95.0%)	3.039855

So the part we are interested in is below:

Hammonasset Beach State Park

Column1		*Column2*	
Males		Females	
Mean	13.81818	Mean	11.09091
Standard Error	2.620111	Standard Error	2.273995
Median	7	Median	4.5
Mode	6	Mode	0
Standard Deviation	17.37985	Standard Deviation	15.08398
Sample Variance	302.0592	Sample Variance	227.5264
Kurtosis	7.318077	Kurtosis	5.408033
Skewness	2.472098	Skewness	2.203718
Range	86	Range	71
Minimum	0	Minimum	0
Maximum	86	Maximum	71
Sum	608	Sum	488
Count	44	Count	44
Confidence Level(95.0%)	5.283957	Confidence Level (95.0%)	4.585949

All the sites

Column1		*Column2*	
Males		Females	
Mean	15.306818	Mean	11.02273
Standard Error	1.9357063	Standard Error	1.529403
Median	9	Median	5
Mode	1	Mode	2
Standard Deviation	18.158534	Standard Deviation	14.34707
Sample Variance	329.73237	Sample Variance	205.8386
Kurtosis	5.9286218	Kurtosis	5.941884
Skewness	2.2681162	Skewness	2.288855
Range	92	Range	71
Minimum	0	Minimum	0
Maximum	92	Maximum	71
Sum	1347	Sum	970
Count	88	Count	88
Confidence Level (95.0%)	3.8474258	Confidence Level (95.0%)	3.039855

If we want to compare these two sets of crab populations we first have to compare males to females in each population.

Hammonasset

Stat	Males	Females
Sum	608	488
Average	13.81	11.09
Sample Variance	302.05	227.52
Standard Deviation	17.37	15.08
N	44	44
Confident level 95%	5.28	4.58

Sample 1 Males

Average -/+95% Confident level = Average will be between 8.61 and 19.09

Sample 2 Females

Average -/+ 95% Confident level = Average will be between 6.51 and 15.67

Now, let's do the other collection set: All sites

Stat	Males 1	Females 2
Sum	1347	970
Average	15.30	11.02
Sample Variance	329.73	205.83
Standard Deviation	18.15	14.34
N	88	88
Confident level 95%	3.84	3.03

Sample 2: Males

Average -/+95% Confident level = Average will be between 11.46 and 19.14

Sample 2: Females

Average -/+ 95% Confident level = Average will be between 7.99 and 14.05

Comparison between the two sets

Sample #	Low M	High M	Low F	High F
1	8.61	19.09	6.51	15.67
2	11.46	19.14	7.99	14.05
% difference between sample 1 and sample 2	75.1	99.7	81.4	89.6

So after all this statistical work is done, what are our conclusions? What have we learned and how can we use it?

A simple statement will help, such as:

- In Hammonasset there are fewer crabs per ¼ of a square meter on average than when considering all sites.
- There are fewer females than males.
- The number of females crabs deviates less at each site than males.
- If we went to a beach in Connecticut and collected Japanese crabs we should expect to find on average between 11 and 19 males and between 7 and 14 females per 1/4 square meter..

My first recommendation would be to go over all these numbers with your students and ask them to run the statistics using their computers and Excel®. Can they come up with more conclusions about the two crab populations? Learning to write a "Conclusions" section for this experiment is very important.
Your next step will be to find an organism and conduct some population studies. You can do this in nature with trees, insects, etc. I have found that one can buy shells in large quantities and conduct a variety of experiments with them.
Two shops that I recommend are Shell World and Shell Warehouse in the Florida Keys. The variety and number of different shells they sell are staggering. There are a large number of other stores on the web. Shells are cheap.

Chapter 4

Name: Population Study of the Rocky Shore Crabs

Activity: Collecting, measuring and describing organisms found in rocky shores.

Level: 1

Developers: Alberto F. Mimo and Anna M. Jalowska

Site: Outdoor

Introduction

An invasive species is a species that is not native to a specific location and has a tendency to spread, taking over habitats of native species, causing damage to the environment, economy or human health. For example, the Asian shore crab (*Hemigrapsus sanguineus*), the western Pacific brachyuran crab, has invaded the coast from southern Maine to North Carolina. The first reported Asian shore crab on the Atlantic East Coast was identified in September 1988 in Townsends Inlet, NJ (McDermott et al., 1991). *H. sanguineus* came to the U.S. East coast in a ship's ballast water. *H. sanguineus* is indigenous to western Pacific regions from Russia, along with the Korean and Chinese coasts, to Hong Kong, and the Japanese archipelago.

The Asian shore crab has a square-shaped shell with three spines on each side of the carapace. The carapace's color ranges from green to purple to orange-brown to red. *H. sanguineus* has characteristic light and dark bands on its legs and red spots on its claws. This species is small with the maximum carapace width ranging from 34 mm (female) to 51 mm (male). Male crabs have a distinctive fleshy, small, hard, bulb-like structure at the base of the moveable finger on the claws. Usually, the male claws are bigger than the female claws.

Male

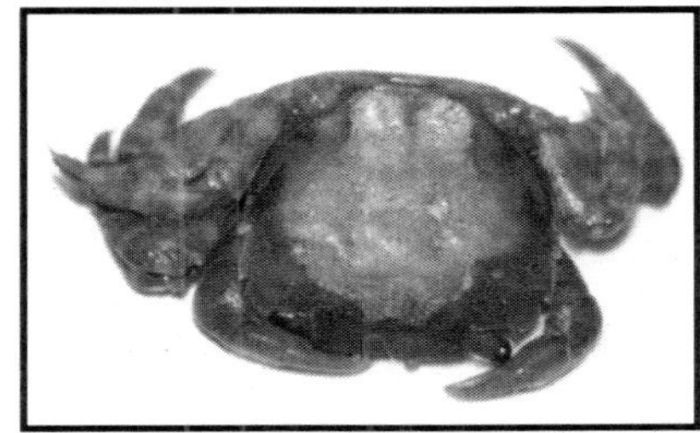

Female

The Asian shore crab is found in rocky intertidal areas where substrates range from large boulders to rocks, cobbles, and broken shells. Some crabs spend time in the subtidal area, especially during the winter. *H. sanguineus* may compete with larger species, like the blue crab, rock crab, lobster, and the non-native green crab. Studies show that the numbers of shore crabs are steadily increasing while native crab populations are declining. *H. sanguineus* poses a threat to coastline ecosystems and aquaculture operations. It an is an opportunistic omnivore; it feeds on a variety of resident organisms, including macroalgae, salt marsh grasses, larval and juvenile fish, and small invertebrates such as amphipods, gastropods, bivalves, barnacles (*Semibalanus balanoide*), and bristleworms (*Polychaetes*). Laboratory studies have shown that the Asian shore crab readily consumes three species of commercial shellfish, the common mussel (*Mytilus edulis*), the soft-shelled clam (*Mya arenaria*), and the eastern oyster (*Crassostrea virginica*). The crab can consume eleven to twelve mussels daily.

Their breeding season (spawning season) is from May to September (twice the duration of native crabs). Eggs hatch within just fourteen days. Females may become ovigerous at about 12 mm, and several broods are produced in a year. The zoea stage of the crab is highly resistant to temperature and salinity changes. Zoea development ranges from sixteen days at 25 °C to fifty-five days at 15 °C. Zoeae can tolerate salinity of 15 ppt at 25 °C, although they require at least 20 ppt salinity at lower temperatures. The megalopa stage is less resistant to wide ranges of temperature and salinity and requires high salinity. Under these conditions, megalopae molt to the first juvenile stage in

Zoea- *H. Sanguineus*

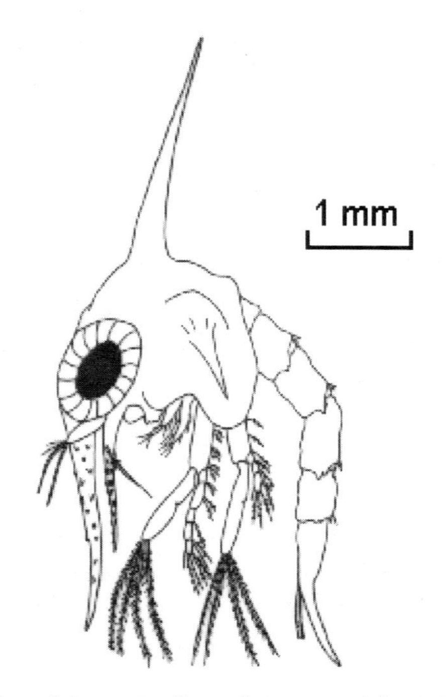

Megalopa- *H. Sanguineus*

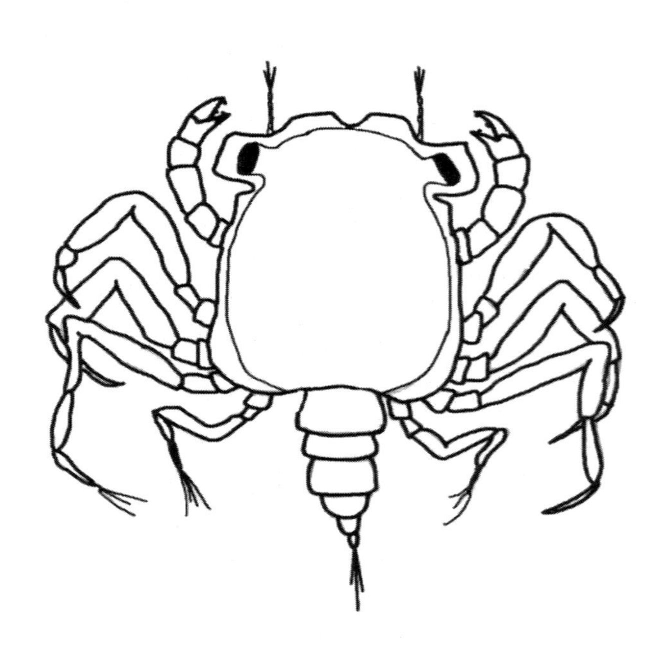

about twenty-five days post-hatching, and in the next thirty-five days to newly metamorphosed crabs. Crabs grow 8 percent of their body mass per day (Epifanio et al., 1998), that is, they attain 112 percent of their body mass in just fourteen days.

All crabs have one pair of **chelipeds** and four pairs of **walking legs**. Chelipeds are used for holding and carrying food, digging, cracking open shells, and warding off would-be attackers. The hard cover or exoskeleton is called the **carapace.** It protects the internal organs: the head, thorax, and gills.

Visible on the underside of a crab are the **mouthparts** and the **abdomen**. The **gills** through which the animal obtains oxygen cannot be seen. They are soft structures under the side of the carapace. The **eyes**, which protrude from the front of the carapace, are on the ends of short stalks. The mouthparts are a series of pairs of short legs, specialized to manipulate and chew food.

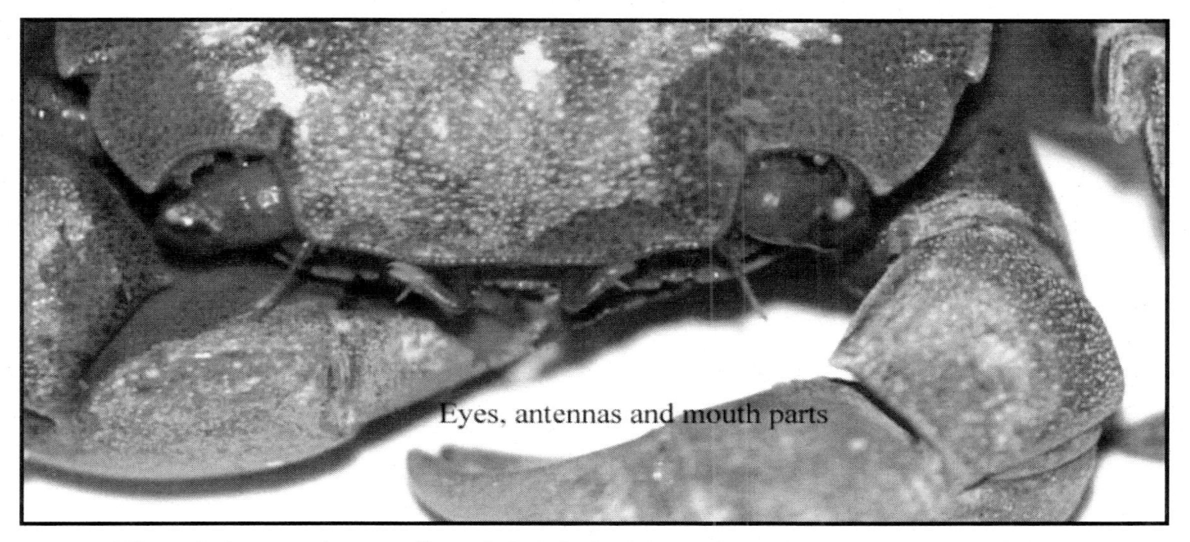

Eyes, antennas and mouth parts

The abdomen is small and tightly held against the underside of the body. Like all crustaceans, the sexes are separate and the size of the abdomen distinguishes them: in males, it is triangular and inset into the underside, while in females it is wide, round and most obvious when eggs are being carried.

Male abdomen

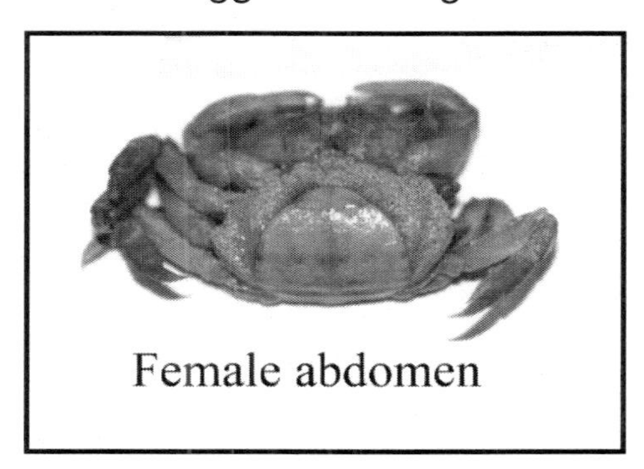

Female abdomen

Crabs are capable of surviving in very cold temperatures in a state of stasis and resume rapid activity if the weather warms up.

Typical Shoreline Transect and Habitat

Tools: Tide charts; GPS; 1 square meter quadrats made with PVC pipe (transects); measuring tape; digital camera; 10 small size buckets (one for every two students); one large bucket; calipers; timers (one for every two students); water quality testing kits (dissolved oxygen, temperature, pH, nitrates, phosphates, salinity).

Objectives

The main objective of this activity is to obtain a clear and concise profile of the rocky intertidal zone. Organisms will be found according to the site location, availability of species, actual proximity to the water's edge, and the substrate and physical features of each plot. Sample inventories and data for this activity can be found at www.northeastnaturalist.com.

Method

Site selection

Sites can be selected by using Google Earth and should be 1) located within a state or municipal property; 2) designated as open space; 3) accessible to school students and with bus parking; 4) must have a rocky intertidal zone; 5) intertidal zone must be accessible during the site visit, best at low tide (check tidal charts).

Capturing and Handling Crabs

H. sanguineus can be aggressive. As a rule, the larger the claw the more powerful it is and the more the required caution should be taken while handling. The best way to capture them is to pin the crab's abdomen with the fingers of one hand (thumb and forefinger), and lay it on your palm with its abdomen up.

Procedure

1. At each research visit, samples should be collected and measurements should be conducted at ten random study plots (transects), each measuring 1 m^2. All plots will be located within the intertidal zone, starting with areas closer to the water line and moving up as the tide arrives.
2. Measure and record total dissolved oxygen, salinity, temperature, and pH readings will be taken at the water's edge.
3. A digital camera will be used to take a picture of each plot to study the substrate.
4. Place a transect at random, and measure the distance to the water's edge at low tide.
5. Record the substrate composition: rock size and shape, any empty open space, and seaweed cover; Percentage of boulders, rocks, gravel, and sand;

Shapes of rocks will be studied by comparing the overall quantity of angularly shaped rocks versus round ones.

6. While turning over rocks and boulders, do so with care,so as not to crush any organisms beneath them (barnacles, mussels, etc.).
7. While removing organisms from the transect, try to replace any overturned rocks, stones, and, later, organisms in their original position.
8. Put all the crabs in separate buckets, one bucket per transect. Buckets can be small.
9. Take one crab at a time; using calipers, measure the carapace of the crab (widest part); identify the sex of the crab (Male, Female, Ov. Female - ovigerous females, Juv - juveniles of size below 9mm), Record the data on a field sheet or research data book.
10. Keep organisms from each site separate, and do not release crabs until the end of the research.
11. Keep crabs cool and moisturized (e.g., cover with seaweed).
12. All organisms will be identified by students according to the genus/species level by using appropriate identification keys.
13. Return crabs to the habitat in which you found them at the end of the activity.

Questions

1. Describe an ecosystem where *H. sanguineus* lives (include abiotic information, substrate, fauna, and flora).
2. Find food chains in the described ecosystem (biophages, saprophages).
3. Analyze: rock size and shape, stone density; site's slope; Tides (living in the intertidal zone, survival under potentially extreme conditions like temperature and dissolved oxygen fluctuation, water loss, wave stress, interruption in feeding); and find the limiting factor for *H. sanguineus*.
4. Compare the diversity of the sites (charts, analyses).
5. Analyze the crab's movements within the intertidal zone.
6. How do crabs change in size within an intertidal zone (charts, analyses)?
7. Compare male and female sizes in the intertidal zone (charts, analyses).
8. Find the correlation between the average abundance of crabs per day, temperature, salinity, dissolved oxygen, and slope (histogram, analyses).
9. Would *H. sanguineus* prefer large, round stones and why?
10. Which species control the communities on these particular sites?
11. How can one complete a census of the crab population?
12. How can we use average populations to study density vs area?
13. Availability of shelter and factors correlated with tidal height influence the density of *H. sanguineus*.

POPULATION STUDY OF THE ROCKY SHORE CRABS. DATA COLLECTION SHEETS

Research Site. Town:

____

School Name:

____

Teacher's Name:

____

High Tide: ______________ Low Tide: ______________

Latitude: ______________ Longitude: ______________

Total number of transects:

____

Slope: ___________

	Beginning of sampling	End of sampling
Wind Speed	0 1 2 3	0 1 2 3
Wind Direction	NE E SE S NONE	NE E SE S NONE
Cloud Cover	0 1 2 3	0 1 2 3
Air Temp (°C)*		
H_2O Temp (°C)*		
Salinity (ppt)*		
Dissolved oxygen (mg/L)*		
pH*		

*average from three readings

Transect Distance to Low Tide Mark: __________

Picture # ______Transect #__________

Latitude: _______________ Longitude: _________________

Research Team Name: ____

Crab Number	Size (mm)	Male (♂), Female (♀), Ov. Female, juv.)	Observations (species, condition)

Summary:

Total Crabs: __________

Males: __________

Females: __________

Mean size: _______________

Male mean: _______________

Female mean: _______________

Standard Deviation: __________

OBSERVATIONS

Substrate:

Plants:

Animals:

Chapter 5

Name: Using Sweep Nets and Collecting Insects Calculate a Food Availability Index for Birds

Activity: Collecting shore insects to study the availability of food for birds. This study can also be completed in your school.

Level: 1

Developer: Alberto F. Mimo

Site: Outdoors

CLIMATE CHANGE PROJECT

Food Availability for Birds Activity

Habitat Assessment

This is a map of Sherwood Island State Park. The park has many habitat types, Forest, Shrub Lands, Grassy Areas, Saltmarshes, and Coastal Habitats such as Sandy Beaches and Sand Dunes.

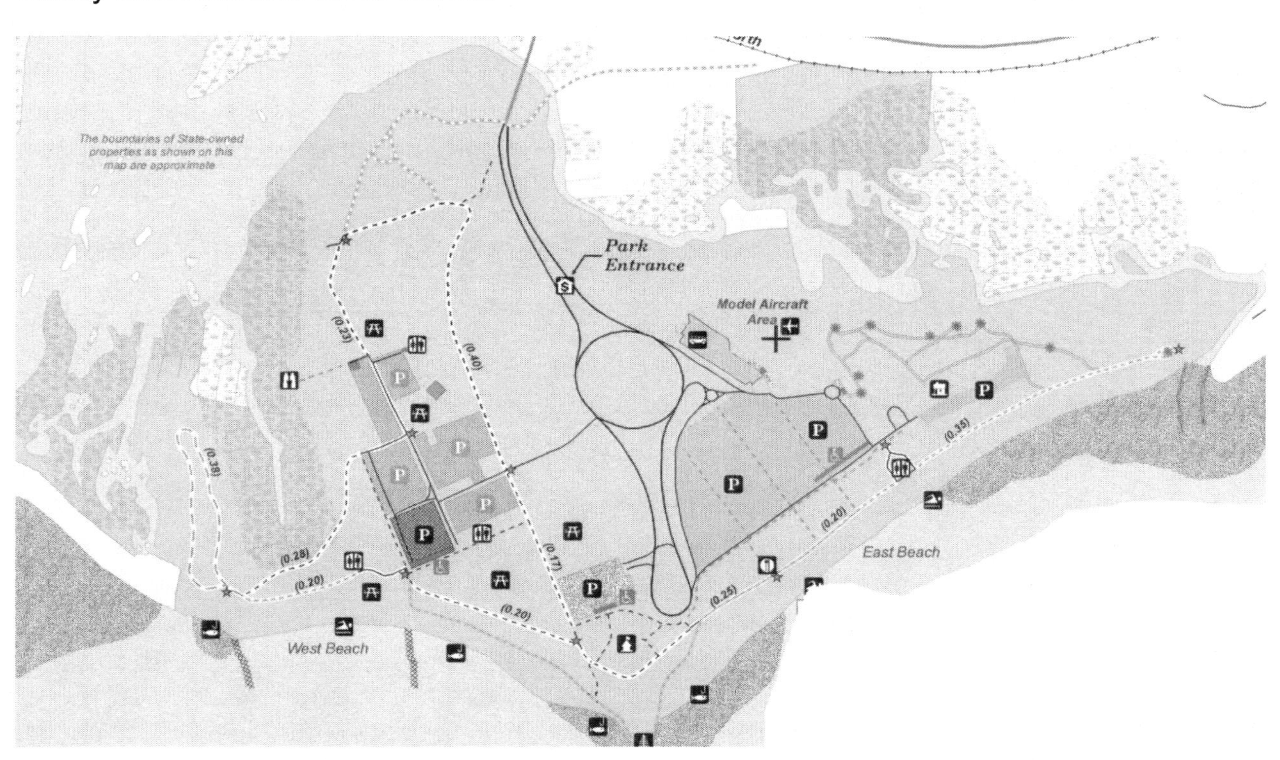

- Using your pencil circle the area where you are standing now. You can make a map of any location and follow the same directions.

Inhabitants of Grass Lands

GRASSES are primitive plants found everywhere on the earth. At home, our lawns are made up of grasses, but in the wild grasses grow in many places such as saltmarshes, prairies, waste areas, river margins, and many other locations. They are a primary source of food for birds and mammals. Grasses are also places for birds to hide and nest.

Grassy areas are rich in ARTHROPODS. Arthropods are animals without a back bone that have jointed legs. So, bugs or insects, spiders, crustaceans, millipedes, and centipedes are examples of ARTHROPODS.

The larger the variety of plants in grasslands the greater the diversity of plant species, which will increase the variety of arthropods and a greater menu of foods for birds.

Picture of Chinese Praying Mantis by Rebecca Belanger

The Praying Mantis is an insect that has 6 legs.

Spiders are arachnids and have 8 legs.

Pill Bugs are isopods which are crustaceans and have two pairs of antennae and 7 pairs of appendages.

Millipedes have two pairs of legs on each segment.

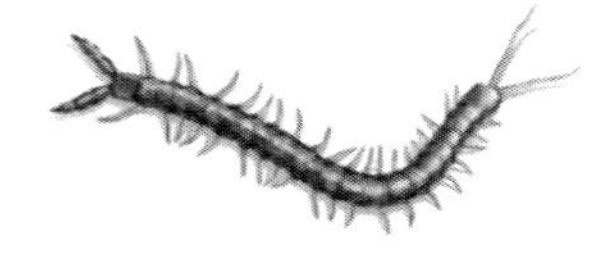

Centipedes have one pair of legs on each segment.

Insects and other Arthropods are the largest group of animals on this planet. They eat plants and other animals, and they are eaten by a number of other animals including BIRDS. Arthropods are a major source of PROTEIN for birds, and an essential part of their diet particularly when birds are having babies in the spring.

Birds' Diet and Timing

Grassland insects hatch at different times during the year. An early hatch happens in June when birds are back from the south but as the weather becomes warmer with climate change June hatches start to happen earlier. Most birds migrate north starting in April but as the weather warms up from climate change they are migrating earlier and the insects are not there yet for them to eat.

Smaller insects, such as ants, and leafhoppers, are good food sources but birds need to eat many of them to make a meal. Large insects such as dragonflies, beetles, and grasshoppers are better sources of protein.

So, do we have an appropriate amount of insects for the birds to eat?

Activity

Invertebrate Availability to Birds
Background Information

Anthropods (organisms with a hard exoskeleton and jointed legs that include spiders, insects, millipedes, centipedes, and crustaceans) are an important food source for birds, especially during the spring. As the climate changes, the emergence of insects, spiders, etc. and the arrival time of migrating birds are becoming *desynchronized*: birds are arriving where insects have not yet hatched, so the birds are unable to find rich, protein-based food and are forced to look for other sources.

As the weather warms up, populations of insects (and other organisms eaten by birds) increase in abundance and weight. It is easier for birds to find food as these populations of food grow. As the weather cools insect population declines, birds must migrate earlier or farther south to look for food.

The objective of this activity is to measure the possible availability of food for the birds on the day you are in the field. To do this you will catch insects, spiders, and other arthropods using sweep nets and weigh the catch to come up with an **Invertebrate Availability Index** for that day, at that place.

Supplies:

- 1 sweep net for every 2 students
- 1- gallon plastic container with a perforated lid (such as pantyhose)
- 1 insect killing kit
- 1 white cotton sheet
- 5 Petri dishes
- 1 balance (0.01 gr. accuracy)(insects are very small)
- 1 calculator
- Data sheet
- Hand held weather station

Killing the Insects or Putting Them to Sleep

Insects can be killed using acetone. Acetone, which is used to take off the nail polish enamel on nails, can be found in most pharmacies. You can also buy acetone in the hardware store.

Soak a small cotton swab or a paper napkin in acetone and place it inside the insect jar. Insects will die after a few minutes.

If you do not want to kill the insects, you can put them to sleep using CO_2. You can buy a CO_2 cartridge at any paint ball gun store. Add a good amount of CO_2 inside the jar and in a few minutes the insects will go to sleep. They will start to wake up soon so do your work fast.

Collecting, Identifying and Recording Insects

1. We will divide the Sherwood Island grass areas into three study areas. The front beach, the mowed parking area, and the back marsh.
2. Take temperature, wind speed, and humidity.
3. Using the sweep nets and following the instructions you will complete 5 sweeps in each area.
4. We will place all the insects in a killing jar.
5. Dump the insects onto a white sheet once they are dead.
6. Sort the insects, using tweezers, into groups of look-alikes.
7. Identify and count the insects using the identification key.
8. Place an empty Petri dish on the scale and tare it.
9. Collect all the insects and place them on one or more Petri dishes.

10. Now place the entire catch on the scale and weigh.
11. Divide the weight of the catch by the total number of insects. Then, multiply that number times 100. This number is called the **Invertebrate Availability to Birds Index**.

Content found on nets which include seeds, plant flowers, and arthropods.

Close-up of the content. Most flowers collected are dandelions.

Arthropods found on the net including flies, beetles, and leafhoppers.

EXAMPLE:

Catch for May 12, 2016

78	Leafhoppers
8	Spiders
14	Bees
6	Flies
106	**Arthropods**
0.4 g	Total weight

- .4/106 = .00377
- .00377 x 100 = .377

Therefore, the Invertebrate Availability Index is 0.377

Drawing Conclusions

Discuss the following questions with your students, and the rest of your group:

- What happens to birds with changes in arthropod availability?
- How do changes in weather affect arthropod availability?
- How do seasons impact arthropod availability?
- What happens to birds when arthropod availability increases or decreases?
- How does the arthropod availability affect the arrival and departure times of birds?
- Connect the issues above to Climate Change.

Data Sheet: Invertebrate Availability for Birds (Field)

Using your magnifying glasses or dissecting scope, closely observe and sort by look-alikes. Use the *Bugs* Key to record the common name and the number you found. Also, record a physical description of each look-alike group.

Invertebrate Common Name	Number Caught	Physical Description of the Invertebrates in the Group
Ex. Spiders	*10*	*8 legs, two body parts: head, body (thorax), 8 eyes*

Total number of invertebrates caught: ___________

Total weight of the catch (in grams): ____________

Total weight divided by number of invertebrates = ______________. This is the average weight of each invertebrate

Average weight x100 = _________. This is the Invertebrate Availability Index for the date they were collected

Temperature: ____________

Wind Speed; ____________

Humidity: ____________

Conclusion:

____

____

____

____

Warning: If you are on a beach try not to disturbed the dunes, and be aware of ticks.

BIRDS' DIET INFORMATION

Plovers

Plovers take advantage of any marine or terrestrial arthropods that are found at the beach. Seaweeds and other decaying marine debris will attract flies to the area that can be eaten by plovers.

Photo by Alberto F. Mimo

Warblers

Warblers eat mostly insects that they pick from foliage or capture on short flights or while hovering to reach leaves. Typical prey includes midges, caterpillars, beetles, leafhoppers and other bugs, and wasps.

Photo by Alberto F. Mimo

Swallows

For the most part, swallows are insectivorous, taking flying insects on the wing. Across the whole family, a wide range of insects are taken from most insect groups, but the composition of any one prey type in the diet varies by species and with the time of year. Individual species may be selective; they do not scoop up every insect around them, but instead, select larger prey items than would be expected by random sampling.

Photo by Alberto F. Mimo

This is just a small fraction of what we know about birds' diet. After you look for some birds and have identified them you can search on the web for their diet and find out how important the insects you have caught are to the birds well-being.

(Pictures of the birds were taken at the Yale Peabody Museum in New Haven, CT, with their permission for this publication).

Chapter 6

Name: Surface Area vs. Volume

Activity: How surface area and volume in mammals helps them regulate their internal temperature.

Level: 2

Developer: Alberto F. Mimo

Site: Classroom

Introduction

Nature has its ways! Throughout evolution homoeothermic animals have always found new evolutionary adaptations to keep up with environmental changes and to deal with new problems.

Here we take a look at keeping warm during periods of cold weather. Animals, and especially mammals, have designed many ways to regulate temperature. Poikilothermic organisms regulate their temperature by seeking locations that will warm them up; such as laying down on a rock that is exposed to the sun. Homoeothermic organisms must use their metabolisms to regulate their temperature but they also conserve energy and heat by changing their location, changing their position and, indirectly, by just being large.

Next time you see a film on polar bears look at their behavior. Some times, and especially after active moments, polar bears need to dissipate heat. A large animal can overheat very easily, especially if it is very active.

In this Field Research Methods Book, we will provide you with one activity that will allow you to explore homoeothermic issues and metabolic pathways.

Drawings by Alberto Mimo

SURFACE AREA VS. VOLUME

We will also study the ratio of Volume vs. Surface Area and how this relates to metabolic activities.

Metabolic regulation in Homeotherms is due to a number of factors such as activity, heart rate, size, and environmental temperature. Some mammals such as the mammoth, during the ICE AGE, were able to control heat thanks to their large size. Let's see how size can help regulate internal temperature.

Surface Area
Surface Area, or the skin around any mammal, is the location where heat would be lost. The larger the surface area ratio compared to the volume the animal has, the greater heat loss. Animals that are large lose more heat, but also produce more heat due to their volume, than animals that are small. Small animals produce less heat. Let's take a look at all of this!

Volume
All mammals use their metabolic processes to produce heat. Larger mammals produce more total heat than smaller mammals, as they are larger and the body heats up using energy. This heat is lost through their surface area.

As the ratio of surface area vs. volume changes, the amount of heat generated is larger than the amount of heat lost. Large mammals such as polar bears would have to dissipate heat to prevent becoming overheated. If you observe polar bears or any other large mammal, such as a moose, you will see that they roll on the ice or get in the water to cool off. Being large is good if you have to survive low temperatures. You may even have to cool off.

Watch your dog and see how it will adopt specific postures to retain or dissipate heat. It will roll up into a ball to conserve heat; By doing that it will decrease the surface area to volume ratio. Or it will spread out and lay on the floor to lose heat when hot; By doing that it increases its surface area.

Math

A = Total surface area

V = Actual volume of the animal

S = Side of the cube

w = Width

h = Height

r = Radius

Let's take the area of a cube and look at its formula. **$A = 6(S \times S) = 6S^2$**

Let's take the volume of the same cube and look at its formula **$V = (S \times S \times S) = 6(S^3)$**

If one plots the progression of these two formulas in a graph as S becomes larger one can see that the totals for surface area and volume become farther and farther apart. Their ratio becomes smaller as S becomes larger.

(If you are using beakers, then the formula for their volume and surface area is $V = \pi r^2 h$ and the surface area is $A = 2\pi rh+2\pi r2$. You can ask the students to calculate the ratios for a cylinder.)

A cube

S Value	Surface Area	Volume	Ratios
1	6	1	6
2	24	8	3
3	54	27	2
4	96	64	1.5
5	150	125	1.2
6	216	216	1
7	294	343	.86
8	384	512	.75
9	486	729	.67
10	600	1000	.6
11	726	1331	.55
12	864	1728	.5
13	1014	2197	.46
14	1176	2744	.43
15	1350	3375	.4
16	1536	4096	.38
17	1734	4913	.35
18	1944	5832	.33

Graphing the progressions

The relationship between Surface Area increase and Volume increase

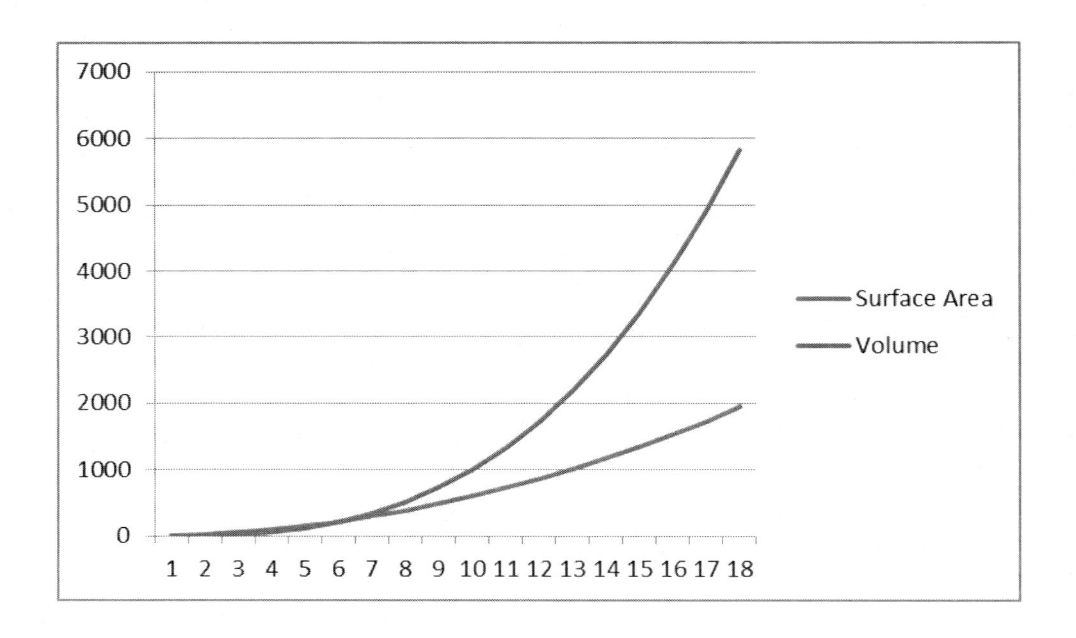

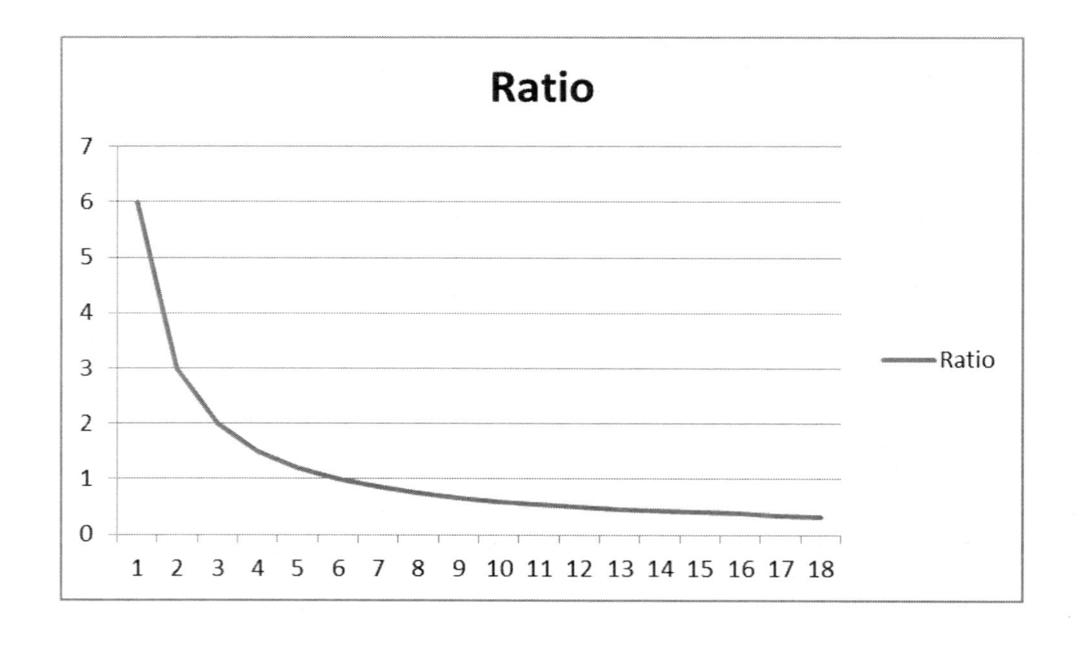

Activity

To prove that the difference in size will make a difference in temperature retention we will use balloons or laboratory beakers, hot water and thermometers. This activity can be done using Vernier equipment or a regular chemistry thermometer. If using beakers use 50, 100, 250 and 500 ml beakers.

Using Vernier equipment you can use the computer to plot the temperature loss and find the equation that best fits your experiment.

1. For this activity, you need 10 balloons or 4 beakers, each of a different size.
2. Fill balloons or beakers with hot water.
3. If using balloons make these balloons different sizes from small to large.
4. Once they are filled, measure the temperature of the water and measure the diameter of the balloon if using balloons. Do this rather fast.
5. You can leave the balloons or beakers at room temperature for 10 minutes. Then measure temperature again.
6. You need to subtract the initial temperature from the final temperature to get the amount of temperature loss in whatever time it took to complete the operation. Be accurate.
7. Graph size of the balloons or beakers vs. temperature loss.

Balloon # Beakers	Diameter	Initial T	Final T	T Loss
1				
2				
3				
4				
5				
6				
7				
8				
9				
10				

Questions:

1. Which balloon or beaker had the greatest decrease in temperature?
2. Which balloon or beaker had the smallest decrease in temperature?
3. Is there a relationship between size and temperature reduction?
4. Can you predict the temperature reduction of a given size of balloon or beaker?
5. What will be the advantage of an animal being a large size?
6. How can an animal retain heat?
7. Is behavior an integral part of retaining heat?

8. Can we describe an activity of small mammals which demonstrates heat control?

9. Are polar bears overheating at any time?

10. Is heart rate dependent on size?

Session Woods Wildlife Management Area, CT (Photo by A. Mimo)

Other Ways to Regulate Temperature

Marine mammals also regulate their temperature thanks to their anatomy. There are two ways that their anatomy helps them: by the production of blubber and by increasing the efficiency of their thermoregulation.

Marine mammals have a fat layer of blubber all around their body. This layer will not only stop cold temperatures from getting in but it also stops heat from getting out.

At the same time, there is a large network of veins and capillaries around the extremities that will help these mammals retain heat. Warm veins from inside the body are side by side with veins coming from superficial or peripheral areas. The warm veins expand and warm the smaller incoming veins.

Blubber Experiment

Place your hand in a container of ice water. Record the length of time you can hold your hand inside. Take your hand out, put a surgical glove on your hand. Smear your hand with Crisco, put another larger glove on top and then place your hand in the ice water. Can you hold your hand inside for a longer period of time?

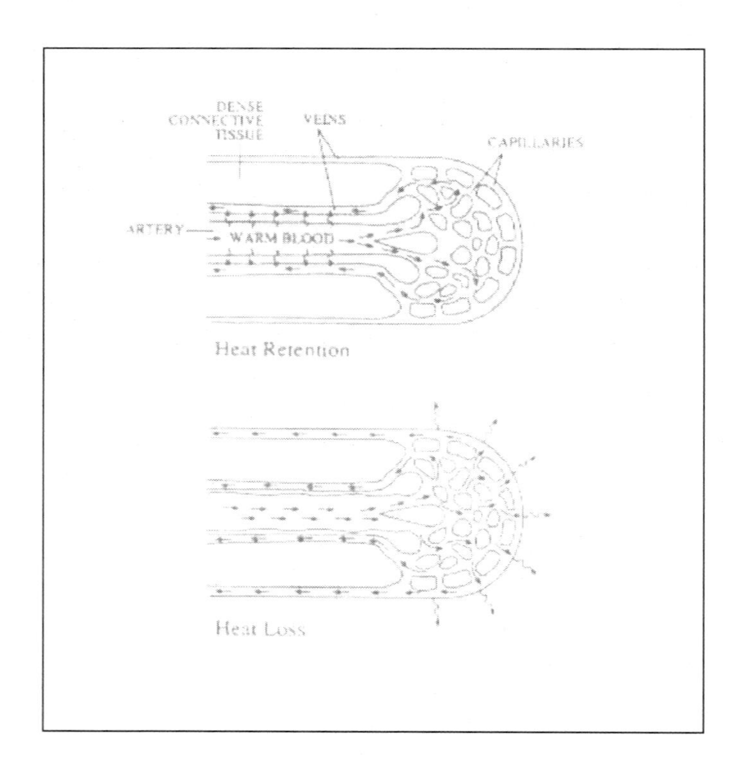

Capillary veins helping to retain heat. Illustration from Scholander, Hock, Waters, Johnson and Irving 1950: Biol. Bull., 99:254

Metabolic Rates and Temperature Regulation in Mammals.

In eutherian mammals, the body temperature lies somewhere between 36° C. and 38° C., in birds it is slightly higher, between 39° C. and 42° C., while in the monotremes and marsupials it is somewhat lower-between 30° C. and 35° C.

Maintenance of a constant body temperature is a neat balance between heat production and heat loss. It demands a sensitive thermostat in the brain. It requires the capacity to use the heat produced as a by-product of metabolism or to output the metabolic energy in accordance with demands. In extreme conditions, animals cannot keep regulating their temperature at the same level and must go into torpidity or hibernation.

Most animals at birth are incapable of maintaining metabolic rates required to keep constant temperatures. Some are completely helpless, such as rats that require 73 days to regulate temperature, as opposed to the caribou that is capable of metabolic regulation at birth.

Here are some metabolic rates for different organisms. As temperature changes, the metabolic rates compensate to keep the organisms within their optimal temperature optimal range.

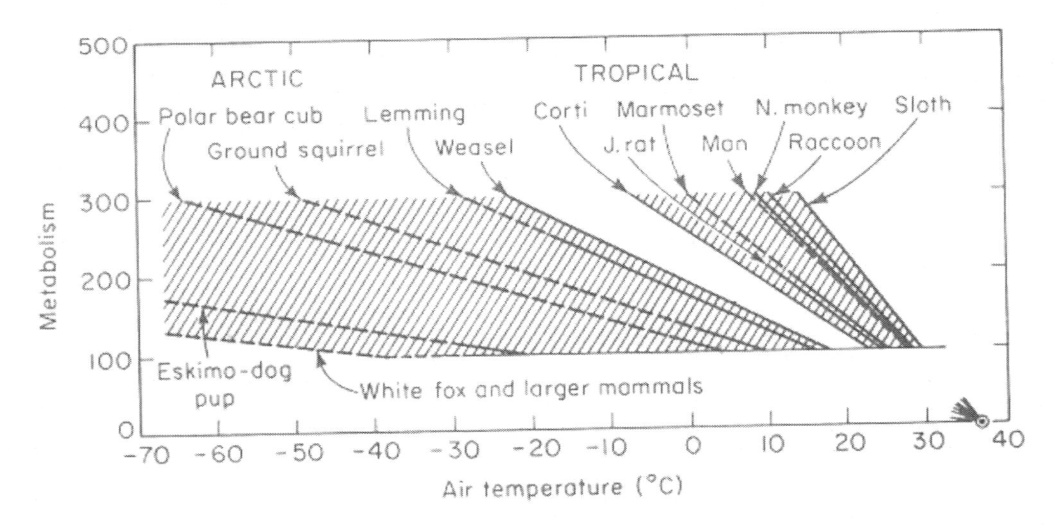

Graph from Scholander, Hock, Waters, Johnson and Irving 1950: Biol. Bull., 99:254

Notice that larger animals require lower metabolic rates. This is because they have a lower ratio of volume to surface area.

What About the Regulation of Temperature in Dinosaurs that Lived in Antarctica?

Dinosaurs could have been cold-blooded, warm-blooded or a stage in-between capable of changing metabolic rates. Studies suggest that there would have been many reasons why they would have fared better as cold-blooded. If dinosaurs were cold-blooded it would make sense as they may have been related to reptiles that are also cold-blooded such as crocodiles. If they were warm-blooded they would have to eat enormous amounts of food which could have depleted the vegetation in an area, or other animal populations, so they would have had to migrate to find food somewhere else. A warm-blooded animal, such as a polar bear or a mouse, has a metabolic rate that is 10 times higher than that of a cold-blooded animal. Birds are born with a very low metabolic rate which increases as the bird grows. That is why, at their early age, they require the protection and warmth of the parents. They develop from cold blooded to warm blooded. Their metabolic rate improves with time. Dinosaurs may have been somewhere in-between cold and warm-blooded depending on their size. Small dinosaurs with as much as 10 times the metabolic rate of a cold-blooded animal; medium size dinosaurs with maybe 5 times the metabolic rate, and large dinosaurs with only 3 or 4 times the metabolic rate of cold-blooded animals. Large dinosaurs such as large sauropods would need to cool frequently by living near water or by being able to lower their metabolism.

We also have to take into consideration that now we have evidence that some dinosaurs were covered with feathers. The feathers would be used to regulate their temperature. The activity of large dinosaurs would decrease during the middle of the day. Large dinosaurs living in cooler areas such as Antarctica would have benefited from regulating their metabolic rate. There have been many dinosaurs discovered who lived in areas close to the South Pole and Australia when the temperature was a little warmer.

Glossary

Capillary veins - small surface veins that carry blood to larger veins that carry blood to the heart.

Eutherian Mammals - a mammal of the major group Eutheria, which includes all the placental animals and excludes the marsupials and monotremes.

Homeostasis - the tendency toward a relatively stable equilibrium between interdependent elements, especially as maintained by physiological processes.

Homoeothermic - an organism that maintains its body temperature at a constant level, usually above that of the environment, by its metabolic activity.

Metabolism - the chemical processes that occur within a living organism in order to maintain life.

Poikilothermic - an organism that cannot regulate its body temperature except by behavioral means such as basking or burrowing.

Thermoregulation -is a process that allows your body to maintain its core internal temperature. All thermoregulation mechanisms are designed to return your body to homeostasis.

Definitions are taken from Google®.

https://www.google.com/?gws_rd=ssl

Chapter 7

Name: Testing for Sodium Chloride in Drinking Water

Activity: Complete a classroom study of sodium and potassiu chloride found in the water supply in your town.

Level: 2

Developer: Alberto F. Mimo

Site: Classroom

How Salty is your water?

Introduction

The two most common salts found in drinking water are sodium chloride (NaCl) and Potassium Chloride (KCl). These two salts are normally found in quantities between 0 and 200 part per million (ppm or mg/l). The acceptable amount according to the Environmental Protection Agency (EPA) should not be greater than 200 ppm.

Winters are hard in New England, and people like to have ice and snow clear from their roads in the winter, consequently sanding and salting of roads is a common practice in every town in this part of the Northeast.

The sand available to do the job is mixed with salt, most of it is Sodium Chloride but if you test it you will find a small amount of Potassium Chloride. The State of Connecticut Department of Transportation sands their roads in the winter with a mix of 7 parts of salt and 2 parts of sand.

Although sanding the roads helps us to clear ice and snow, the salt and sand used is not really good for the environment. Sand increases sedimentation in streams and lakes and salt is a foreign product in fresh water, harmful to freshwater fish and other organisms. Increased doses of salt in drinking water could also be harmful to our health, so a careful monitoring of Sodium Chloride in drinking water is a good idea.

Storage area to keep salt and sand used on the roads.

Collecting the Samples

Any good research project is always dependent on the number of samples taken. Research quality and quantity are often closely related. The other major concern is that care and commitment is taken in the area of Quality Control and Quality Assurance. Careless techniques make poor research.

In order to get a good number of samples, the best method is to involve a group of students to participate in the program, let's say the 6th graders. So your high school students do the testing and mapping and so forth, but the 6th graders collect the samples at each of their homes. By doing this you will get more than 100 samples. Samples should be well distributed within the town and should be random. These samples can be added to a database and by using GIS (Geographic Information Systems), Google Earth or manual mapping you can transfer your results into a visual map that shows the levels of Sodium Chloride in each of the town areas.

People at home either get their water from the town or they have a well. So it will be important to know this information on the source of the drinking water. Also, you want to know where the samples come from. To make this research anonymous do not include the street number, only the name of the street. It will also be extremely important to know that you gave good directions on how to get the sample and any other procedures.

The Letter

The first task is to write a letter directed to the teachers of the entire 6th grade class and to the parents of the students with instructions and information about the project. Teachers should be able to know what will be done, and who is participating. How the project is accomplished and so forth.

Content List for That Letter

- Each 6th grade student should receive a letter for their parents with instructions how to get the samples.
    - The sample should come from the kitchen faucet
    - Water should be run for one minute before taking the sample.
    - The bottle should be rinsed with the sample water three times
    - Each sample of water should be correctly labeled
    - Each parent should fill out the questionnaire
- Each student should get an adhesive label to write the address and date of when the sample was taken.
- Each student should get a form with questions to accompany each sample.
- Each student should get a plastic bottle for the sample. Bottles do not need to be more than 50 ml.
- Bottle, label and questionnaire should be numbered before it is sent out to the 6th grade. Keep all numbers and information on a list so that you can keep track of the procedure.

Questions

- Address (no number)
- Number of people in the household
- Is this water from a well or city water?
- Name of the reservoir where the water comes from

All samples should be tested immediately within the next two days. Keep all samples refrigerated to minimize evaporation. It is best to do this project at the end of the winter season.

Testing the Water

Testing of the water will be performed by using the Chlorides test kit put out by LaMotte Company of Maryland (Kit number 4503-DR-02). See www. LaMotte.com. The price is approximatly $50 for 50 tests. You will need 4 kits.

This kit will test for chlorides within an accuracy of 4 ppm.

The procedure requires a titration and the results can be read directly from the titrator.

There are three steps to this test.

- Filling the test tube
- Adding Chloride Reagent A (three drops)
- Titrating the sample using Chloride Reagent B

To improve efficiency, this entire project can be done as an assembly line. Each student should perform one job. This is the best arrangement.

- Student one – Fills a test tube
- Student two – Adds Reagent A
- Student three – Titrates with Reagent B
- Student four – Writes results in the accompanying questionnaire
- Student five – Enters results to the database to be mapped by other groups of students.

As a QAQC (Quality Control and Quality Assurance) project, a number of the same samples should be brought to another table where they are retested let's say 10% of the samples. In the end, students can do a statistical correlation between each retest (first results to second results) and calculate if the error is within acceptable limits. Statistical Field Techniques can be helpful. You can also include your math teacher.

To complete this project you will need at least 4 test kits, but it will always depend on how many tests you need to perform. Each test kit is good for 50 tests.

- You will need small plastic bottles or vials that hold 50 ml. and close tightly. You will need as many as there are 6th graders.
- The price of the bottles will be between $40 to $80 for 100. The price of test kits will be $50 so 4 kits will be about $200.
- In some cases, you may also be interested in comparing the ratio of Sodium Chloride to Potassium Chloride. You can do that by testing some of the samples for Sodium with test kit number 7791-DR-01. You should have a ratio of no more than 1.2: 1 of potassium to sodium.

Analyzing Your Results

Once you have collected all the data, you need to protect your results. People do not like having their address published with information regarding the test results. The best way to publish this information is by using a map of the town and color coding areas by range. Use street names to help identify areas.

Your results will range from 0 to 200 ppm. So divide the results into frequencies and map them. Let's say:

- 0-20 Light Blue
- 21-40 Dark Blue
- 41-60 Light Yellow
- 61-80 Dark Yellow
- 81-100 Light Orange
- 101-120 Dark Orange
- 121-140 Light Red
- 141-160 Dark Red
- 161-180 Light Brown
- 181-200 Dark Brown

Hypothetical Sample

Test Range	Frequency	Color
0-20	23	Light Blue
21-40	44	Dark Blue
41-60	20	Light Yellow
61-80	14	Dark Yellow
81-100	12	Light Orange
101-120	5	Dark Orange
121-140	6	Light Red
141-160	3	Dark Red
161-180	5	Light Brown
181-200	12	Dark Brown
TOTAL:	144	

Hypothetical Map

This map of Prospect, Connecticut does not reflect true testing. This is just an example. No testing was done to complete this map.

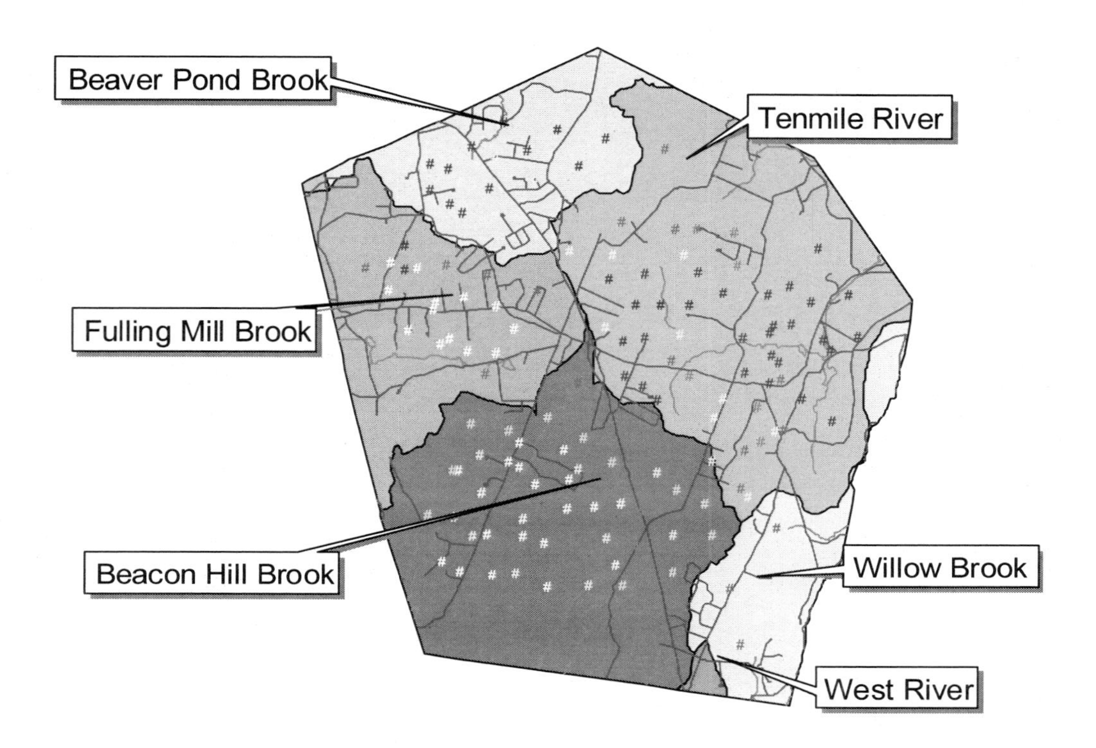

Mapping Techniques

It could be understood that an area with a perimeter of 200 ft. would probably be affected by the action of the salt. Therefore a buffer of 200 ft. can be applied to every one of the points. We can also show the map by making the dots larger as the amount of NaCl is greater.

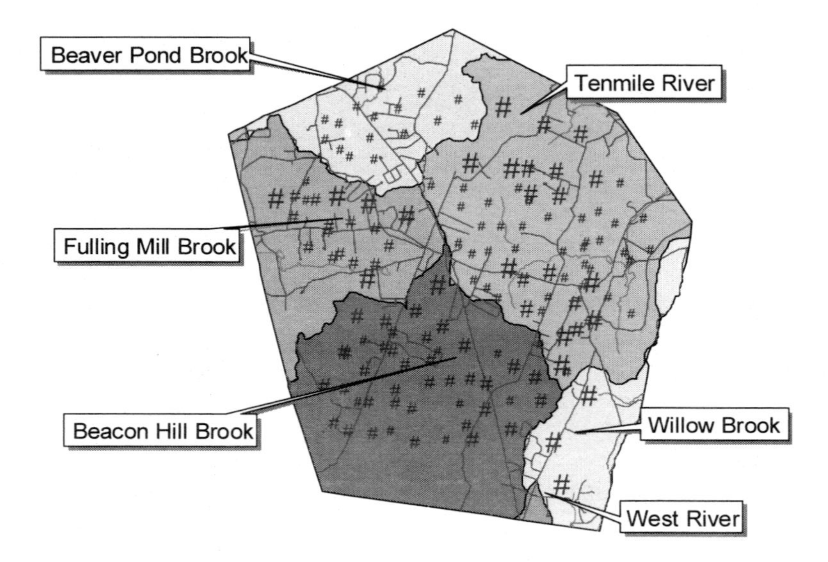

The objective of creating this map is to provide a visual description of the results and show the areas that are most affected.

This project will show the students how a town can be affected by a single pollutant. Although to make this project valid the samples would need to be tested by an authorized laboratory, the experience for the students is valid.

Chapter 8

Name: Saltmarsh Sediment Sample Collection and Analyses

Activity: Geological and biological analyses of saltmarsh-sediment samples that will provide comparison between today and up to 500 years history.

Level: 3

Developers: Alberto F. Mimo and Anna M. Jalowska

Site: Outdoors and Laboratory

Introduction

This research method is divided into three parts:

I. Core collection (field activity).
II. Core processing and analyses of the sediment properties.
III. Core processing and analyses of the sediment biological components.

Sediment is a naturally occurring material broken down by weathering and erosion. With time, sediments travel through river systems to be delivered to the coastal areas and the ocean. On their way to the coast, sediments change their physical and chemical properties. Sediments are composed of inorganic sand and clay as well as organic carbon, and a variety of microscopic organisms, such as diatoms, algae, and foraminifera. Sediment properties can tell us a history of the accumulation site.

The objective of this research method is to take sediment core-samples and analyze their content and composition. The top of the core contains recent sediments. Approximately, one centimeter represents 1 to 10 years of sediment accumulation, so 10 cm from the top can represents 10 to 100 years of sediment history. However, a single weather event, like Hurricane Sandy can leave a distinct layer of 3-10 cm of coarse sediments.

Part I. Core Collection

TOOLS: Cores could be collected using a core sampler, auger or self-made push-core.

Core sampler

Here we present a metal tube with extensions and a hammer attachment (see pictures). To insert the plastic tube in the mud, use the extensions and a hammer system. The plastic core needs to be hammered into the mud, and sample taken by lifting the core cylinder out the sediment floor.

Core sampler with extensions and hammer attachment. The core sample fills the inner part of the tube.

Sample collection

1. Assemble the core sampler.
2. Once the core sampler is ready to be used, walk to the area where you want to get the sediment core. Place the core liner on the surface and hammer it into the sediment.
3. The core liner needs to be capped with a rubber piece to prevent the sample from exiting the liner. The cap produces suction.
4. Loosen the corer by wobbling it and pull it out of the sediments by holding the liner (not the hammer).
5. Disassemble the core from the handles.
6. Open the bottom part and remove the sediment - mark the bottom part of the core.

The core sample may also be collected using an auger. Follow instructional videos and instructions on the web:

https://www.youtube.com/watch?v=upJ03FCE2RI

http://www.organiclawndiy.com/2009/06/make-your-own-soil-sampler.html

Self-made push-core sampler

Core samplers are expensive; however, your hardware store offers cheaper substitutes. For example, the core sampler can be replaced with a PVC, PET or aluminum pipe with Quik caps (pictures).

1. The pipe needs to be pushed into the mud.

2. The Quik cap needs to be put on top of the corer, push it down in the middle and secure it tightly to form the vacuum.
3. Pull the core from the mud and cover the bottom of the core with another Quik cap, the moment the core is out of the sediments.

The best way to transport cores is in vertical position, marking the top and bottom of the sample.

Example of self-made push-core samplers.

Part II. Sediment Properties

TOOLS: Core sample, camera, ruler or measuring tape, spatula/knife, aluminum boats, toaster oven (40°C and 400°C), scale, mortar and pestle, 63-micron and 32-micron sieves, ceramic cups, torch, fire extinguisher.

1. Using the plunger, push the core out of the core liner.
2. Place it in the tray.
3. Photograph the core.
4. Wrap it in aluminum foil.
5. Rinse the core liner and all the parts of the core sampler.

Sample processing

1. Place the ruler along the core. Remember to mark the top and the bottom of the sample.
2. Measure the length of the core and prepare the appropriate number of aluminum boats.
3. Describe the boats and weigh them (record everything on the data sheet).
4. Cut the mud core across its length in half. One-half will be used for sediment properties analyses and the second half for biology analyses.
5. Examine the core for distinct laminations (layers).

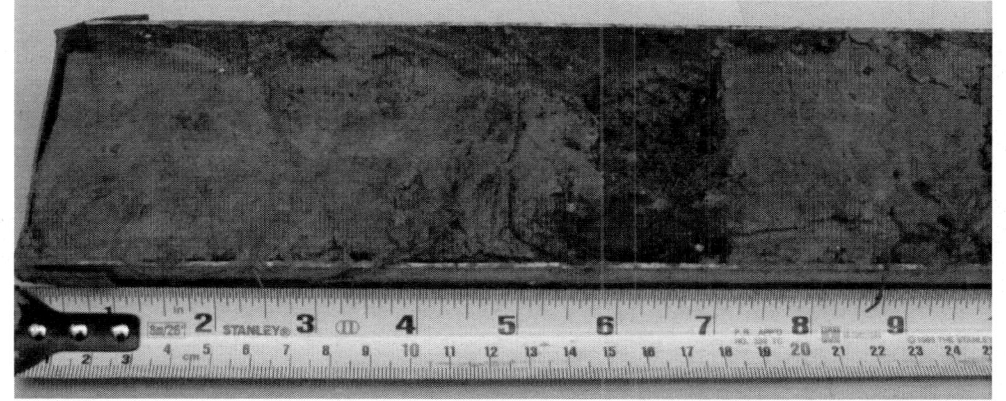

6. Using the knife, cut it into intervals, 1cm will be sufficient for analyses but it is important not to cut through the laminations.
7. Describe and record each of the sample properties (sand/silt/clay, gray/orange/dark, shells/pebbles/organic fibers/ twigs, etc.).
8. Using a piece of paper, draw the core and all the color and texture changes observed. Use a ruler to measure color and texture changes.

Porosity

1. Record the weight of the container in the porosity worksheets (provided).
2. Record the weight of the wet sample.
3. Place aluminum boats with wet samples in the toaster oven and set at BAKE and 200F for 2 hours.
4. If sediment is not dry, add extra baking time.
5. When dry, cool it down and weigh it again. Record the weight.

Lost on Ignition (LOI)

1. Put the sample in the mortar and gently disintegrate the sample.
2. Put it in the boat and weigh it. Record the weight in the LOI worksheet (provided).
3. Set the toaster oven to BROIL and maximum temperature.
4. Put samples in for 1 hour.

5. **NEVER leave the samples! Organics can catch on fire. You should have a fire extinguisher on the site.**
6. Cool the samples and weigh them again.
7. Another way to do the same is to place the sample in a ceramic cup and use a torch for a few minutes to burn the sample. The ceramic cups will be very hot. Wait for 20 to 30 minutes before touching them. Do this outside.
8. You can also ignite the sample by placing the sediment in a ceramic cup and using a small torch, (a small propane tank with a torch at the end). The ceramic cup also gets hot.

Grain Size

1. Prepare the sieves.
2. Weigh each size sample, and record the weight in the Grain Size worksheet (provided).
3. To separate sand from silt and clay use the 63-micron sieve and weigh the sediment left on the sieve. This provides the weight of the sand fraction.
4. To separate silt from clay, put the remaining sediment through the 32-micron sieve. Weigh the sediment left on the sieve. This provides the weight of the silt.
5. Weigh the remaining sediment, which represents the clay fraction of the sample.
6. Calculate percentages.

Analyses

Using a computer, I have prepared an Excel® sheet that makes all the calculations. The Excel® sheet is called LOI. Just make sure you do not erase the calculation areas on the sheet. To use it just add the numbers you will get from the exercise. (You can find all the Excel® files on my website under publications. http://www.northeastnaturalist.com/)

Plot all the results from porosity, LOI and grain size data:

- Do we see differences in % sand along the core?
- Do we see differences in % organic matter in the core?
- Why do they look like a mirror image?
- Why there are wiggles in grain size profiles?
- Can you identify distinct patterns?
- If so, what could the patterns be indicative of?
- Can you identify seasons in the sample?
- Can you identify weather events in the sample?

Porosity worksheet

Sample Name/Depth (cm)	**Container Mass (mg)**	**Wet Sample Mass (mg)**	**Dry Sample Mass (mg)**

LOI worksheet

Sample Name/Depth (cm)	**Container Mass (mg)**	**Dry Sample Mass (mg)**	**Burned Sample Mass (mg)**

Grain Size worksheet

Sample Name/ Depth (cm)	Total Mass (mg)	Sand Mass (mg)	Silt Mass (mg)	Clay Mass (mg)

Part III. Analyzing the Sediment for its Biological Components

Tools: Distilled water, pipette, microscope, microscope slides, Clorox, centrifuge.

Diatoms

Over time diatoms in the water slowly collect in the bottom and mix with the sediment. As sediment accumulates, the diatoms are buried and may be preserved for thousands of years. Diatoms reflect the quality of the water and climatic changes of the past.

Biological Analyses

Subsample the second half of the core in two-centimeter intervals for analyses (2 cm may represent 2 to 200 years of sediment deposition).

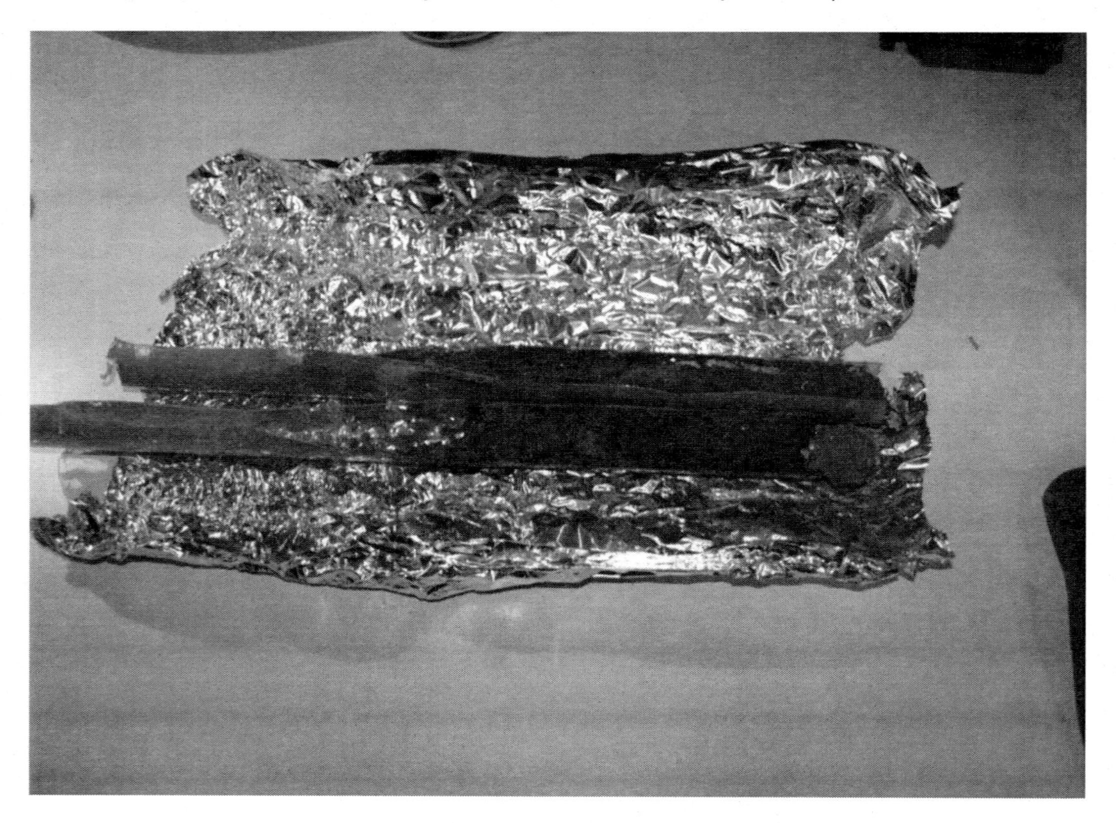

Cross sectional cut of the core to be biologically analized

Diatoms are very durable and may be difficult to identify to genus and species, however, they can be divided into different varieties and analyzed for:

- Taxa diversity
- Abundance
- Biomass

The following information was obtained from the "Diatoms of the United States" website, a very useful resource for diatom identification.

https://westerndiatoms.colorado.edu/

Sample preparation

Without cleaning:

The diatoms can be examined by dissolving the soil sample in the water, but this method would result in unclean diatoms and murky images under the microscope, making the diatom identification challenging.

1. Take the soil sample and place it in a small dish with 5 ml of distilled or clean water, and mix thoroughly.
2. Using a pipette place 1 or 2 drops of the mixture on a microscope slide.
3. Place a cover slip on the slide and place it under minimum magnification on the microscope. Focus on the sample.
4. Bring magnification to X400, adjusting the focus at every change.
5. What you will see is what we will call a "field".

Procedure to clean the diatoms:

1. Take the soil sample and place it in a small dish with about 20 ml of water.
2. Mix the soil sample in the water until it becomes homogeneous.
3. Use a sieve clean the sample of a large debris.
4. The remaining liquid should be a brown homogeneous mixture composed of silt, clay and diatoms. Add 20 to 50 ml of Clorox and set aside for 1 to 2 hours, mixing the sample several times.
5. Use a centrifuge to separate the clay, silt and diatoms from the liquid.
6. Add water to rinse the diatoms. Centrifuge the sample again.
7. Add water to centrifuged sample.
8. Place the sample on a microscope slide and place it under microscope to look for diatoms.

Diatom Morphology

Diatoms are algae with distinctive, transparent cell walls made of silicon **dioxide** hydrated with a small amount of water ($SiO_2 + H_2O$). Silica is the main component of glass and hydrated silica is very like the mineral opal, making these algae, often called "algae in glass houses" more like "algae in opal houses". The diatom's cell wall is called a **frustule** that consists of two **valves**. The silica is impervious; diatoms have evolved elaborate patterns of perforations in their valves to allow nutrient and waste exchange with the environment. These valve patterns can be quite beautiful and helpful in diatom classification. Diatoms grow as single cells or form filaments and simple colonies.

Frustule – The siliceous parts of the diatom cell wall. Composed of the larger epitheca and the smaller hypotheca. The epitheca overlaps the hypotheca similar to a pill box or Petri dish. From Latin for a little piece.

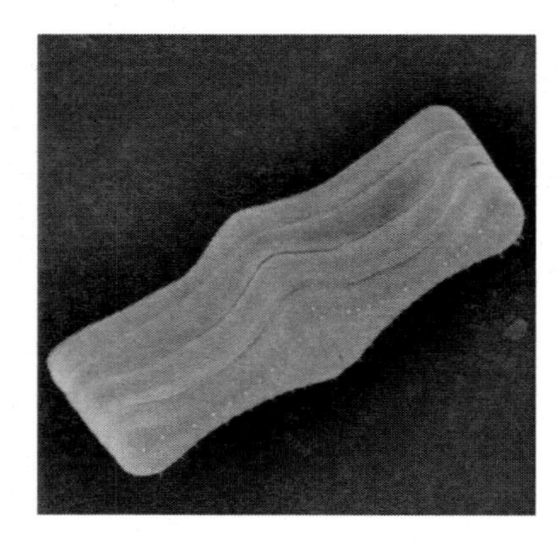

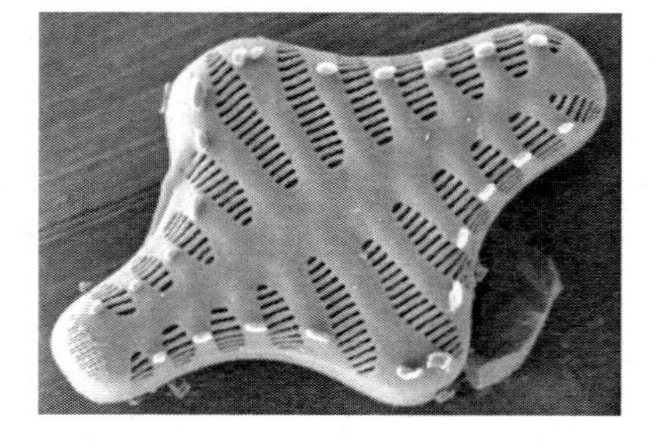

Valve - The siliceous unit that lies at each end of the frustule. The two valves are linked by the girdle bands.

Raphe - One or two slits, or fissures, through the valve face of monoraphid and biraphid diatoms. If two slits are present, each is called a branch of the raphe. Branches may be separated by a silica thickening called the central nodule. Raphe position may be (1) axial, along the apical axis; (2) eccentric, along one margin; or (3) circumferential, around the whole margin of the valve.

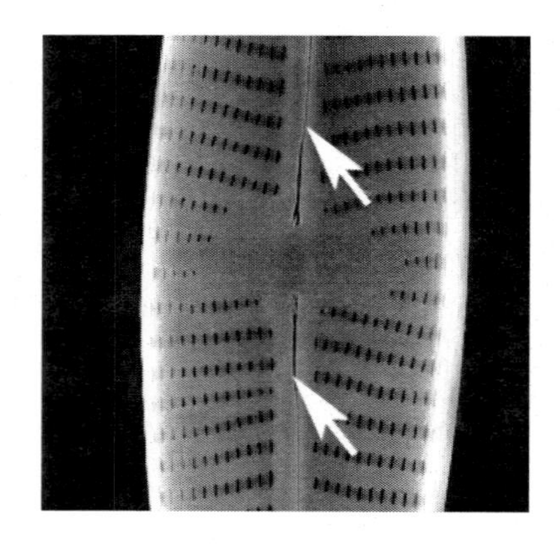

Diatom Identification

There are nine varieties of Diatoms based on their shape: This section has been obtained from the "Diatoms of North America" Website with their permission, and it may be helpful to navigate through their website.

Centric

- Valves with radial symmetry (symmetric about a point),
- Cells lack a raphe system and lack significant motility,
- Cells may possess fultoportulae (strutted processes) and rimoportulae (labiate processes),
- Sexual reproduction is oogamous.

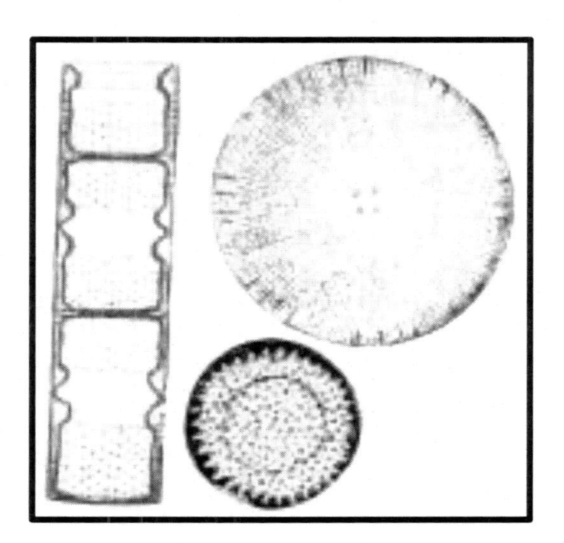

https://westerndiatoms.colorado.edu/taxa/_morphology_guide/centric

Araphid

- Valves with bilateral symmetry (symmetric about a line),
- Cells lack a raphe system and lack significant motility,
- Rimoportulae (labiate process) may be present.

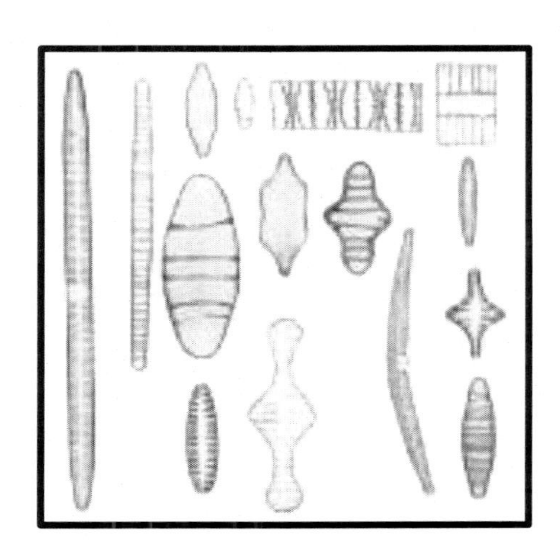

https://westerndiatoms.colorado.edu/taxa/_morphology_guide/araphid

Eunotioid

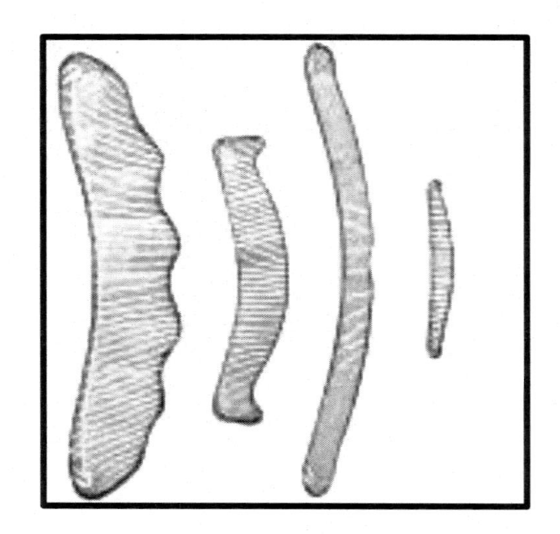

- Valves with bilateral symmetry (symmetric about a line),
- Valves often asymmetrical to the apical axis,
- Raphe system is short and provides weak motility,
- Raphe located on valve mantle and face,
- Cells may possess two or more rimoportulae (labiate processes).

https://westerndiatoms.colorado.edu/taxa/_morphology_guide/eunotioid

Symmetrical biraphid

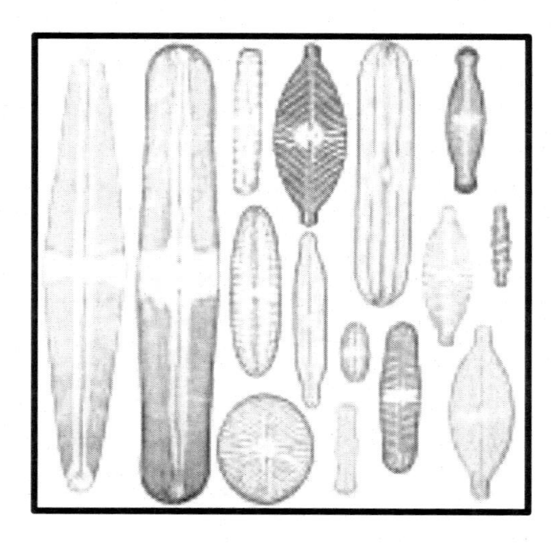

- Valves with bilateral symmetry (symmetric about a line),
- Valves symmetrical to both apical and transapical axis,
- Raphe system well developed and cells may be highly motile,
- This group has the greatest diversity among the freshwater diatoms.

https://westerndiatoms.colorado.edu/taxa/_morphology_guide/symmetrical_biraphid

Monoraphid

- Valves with bilateral symmetry (symmetric about a line),
- Raphe system present on one valve (raphe valve),
- Raphe system absent on one valve (rapheless valve),
- Heterovalvar ornamentation.

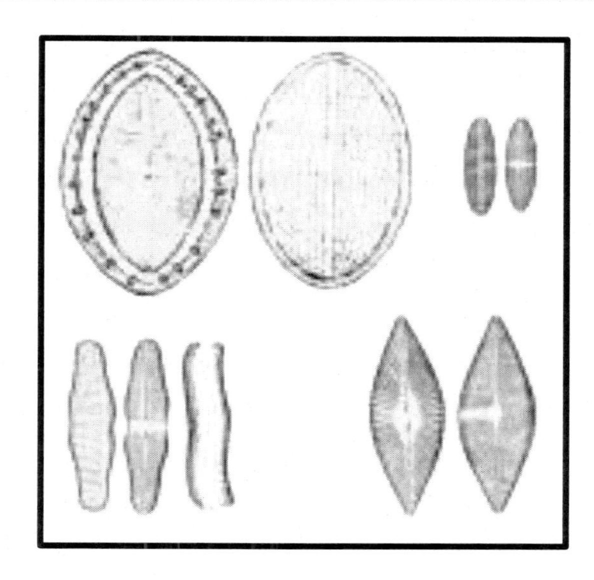

https://westerndiatoms.colorado.edu/taxa/_morphology_guide/monoraphid

Asymetrical biraphid

- Valves asymmetrical to apical axis OR asymmetrical to the transapical axis, or both,
- Raphe system well developed,
- Some genera possess apical porefields that secrete mucilaginous stalks,
- Other genera secrete mucilagous tubes.

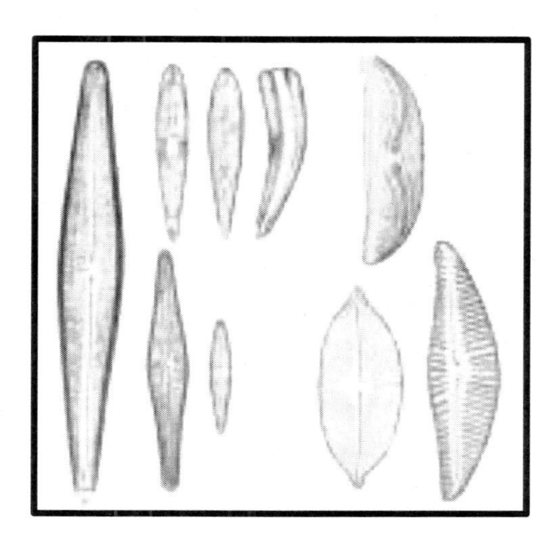

https://westerndiatoms.colorado.edu/taxa/_morphology_guide/asymmetrical_biraphid

Epithemioid

- Valves with bilateral symmetry (symmetric about a line),
- Valves asymmetrical to apical axis,
- Raphe system well developed and enclosed within a canal,
- Raphe system positioned near the valve margin.

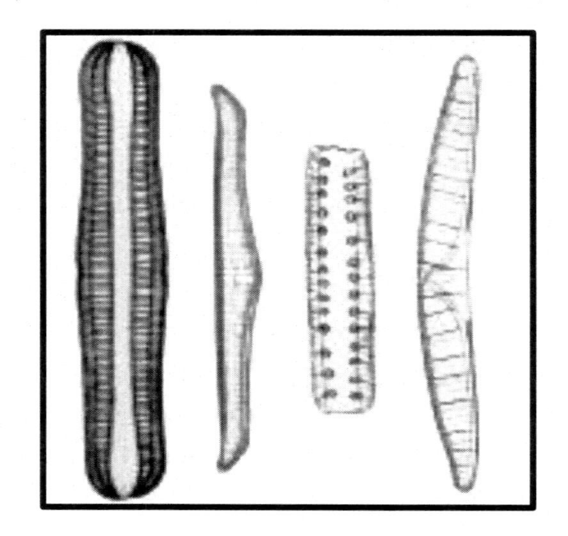

https://westerndiatoms.colorado.edu/taxa/_morphology_guide/epithemioid

Nitzchioid

- Valves with bilateral symmetry (symmetric about a line)
- Valves usually symmetrical to both apical and transapical axes.
- Raphid system well developed, and positioned near the valve margin.
- Raphe is enclosed within a canal and may be raised onto a keel.

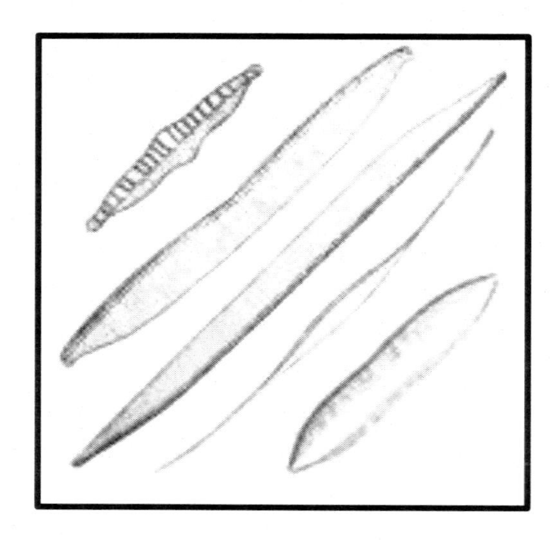

https://westerndiatoms.colorado.edu/taxa/_morphology_guide/nitzschioid

Surirelloid

- Valves with bilateral symmetry (symmetric about a line),
- Raphe system extremely well developed and enclosed within a canal,
- Raphe positioned around the entire valve margin and raised onto a keel.

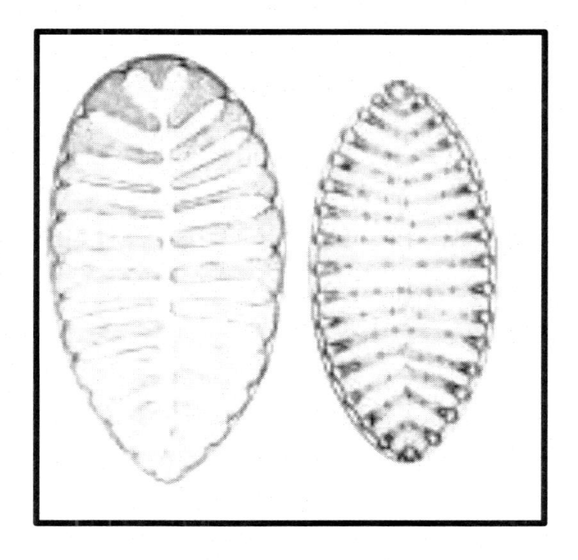

https://westerndiatoms.colorado.edu/taxa/_morphology_guide/surirelloid

Calculating Taxa Richness

Taxa Richness is simply the total number of shapes you encounter. This will be a number between 0 and 9

Taxa Diversity

1. Take a good look at the "field" and count the number of different shapes you see. (type of diatoms by morphology)
2. Group the different shapes and count their numbers. Look at the"Diatoms of North America" website to help understand. How many centrics, how many eunotids etc.
3. Add the numbers to get a total.
4. Do the same for an additional 9 fields. Fill out the worksheet.

EXAMPLE

Field	Type	Count	Total	Taxa Diversity	Biomass
1	Centric	4			
	Araphid	15			
	Surireloid	2			
	Assy. biraphid	1			
	4		22	0.48	23 %

Taxa Diversity Calculation

Formula $TD = \sum n(n-1) / N(N-1)$

Where n = # organisms per taxa and N = Total number of organisms

Sample:

4(4-1)+2(2-1)+ 15 (15-1) + 2 (2-1) + 1(1-1) / 22 (22-1) =
(4x3)+(2x1)+(15x14)+(2x1)+(1x0)/(22x21)=

12+2+210+2+0/462 = 224/462 = 0.48

There is a 48 % probability that on your next search you will find taxa that you have already seen. The lower the TD number the higher the diversity.

Biomass

Biomass of the sample is the loss of weight of the sample once it has been burned. This should be done after drying the sample. This should be done by the teachers as a demonstration. The sample and the cup get really hot. See the section on Loss on Ignition.

1. You will need a torch, a ceramic cup,a small sample, and a balance.
2. Weigh the ceramic cup, and tare it.
3. Add the sample of sediment and weight the sample.
4. Torch the sample. Be careful. Place it on a metal tray or cookie sheet.
5. Leave it on the tray until it is fully cool.
6. Weigh the sample again. The difference in weight is how much carbon has burned out.
7. Give the answer in percentage mass.

Percentage mass = Burned mass x 100/original weight

Biological analyses worksheet

Field	Type	Count	Total	Taxa Diversity	Biomass
1	Centric				
	Araphid				
	Eunotioid				
	Symmetrical Biraphid				
	Monoraphid				
	Asymmetrical Biraphid				
	Epithemioid				
	Nitzchioid				
	Surirelloid				
Totals					

Field	Type	Count	Total	Taxa Diversity	Biomass
2	Centric				
	Araphid				
	Eunotioid				
	Symmetrical Biraphid				
	Monoraphid				
	Asymmetrical Biraphid				
	Epithemioid				
	Nitzchioid				
	Surirelloid				
Totals					

Field	Type	Count	Total	Taxa Diversity	Biomass
3	Centric				
	Araphid				
	Eunotioid				
	Symmetrical Biraphid				
	Monoraphid				
	Asymmetrical Biraphid				
	Epithemioid				
	Nitzchioid				
	Surirelloid				
Totals					

Field	Type	Count	Total	Taxa Diversity	Biomass
4	Centric				
	Araphid				
	Eunotioid				
	Symmetrical Biraphid				
	Monoraphid				
	Asymmetrical Biraphid				
	Epithemioid				
	Nitzchioid				
	Surirelloid				
Totals					

Field	Type	Count	Total	Taxa Diversity	Biomass
5	Centric				
	Araphid				
	Eunotioid				
	Symmetrical Biraphid				
	Monoraphid				
	Asymmetrical Biraphid				
	Epithemioid				
	Nitzchioid				
	Surirelloid				
Totals					

Field	Type	Count	Total	Taxa Diversity	Biomass
6	Centric				
	Araphid				
	Eunotioid				
	Symmetrical Biraphid				
	Monoraphid				
	Asymmetrical Biraphid				
	Epithemioid				
	Nitzchioid				

	Surirelloid
Totals	

Field	Type	Count	Total	Taxa Diversity	Biomass
7	Centric				
	Araphid				
	Eunotioid				
	Symmetrical Biraphid				
	Monoraphid				
	Asymmetrical Biraphid				
	Epithemioid				
	Nitzchioid				
	Surirelloid				
Totals					

Field	Type	Count	Total	Taxa Diversity	Biomass
8	Centric				
	Araphid				
	Eunotioid				
	Symmetrical Biraphid				
	Monoraphid				
	Asymmetrical Biraphid				
	Epithemioid				
	Nitzchioid				
	Surirelloid				
Totals					

Field	Type	Count	Total	Taxa Diversity	Biomass
9	Centric				
	Araphid				
	Eunotioid				
	Symmetrical Biraphid				
	Monoraphid				
	Asymmetrical Biraphid				

Field	Type	Count	Total	Taxa Diversity	Biomass
	Epithemioid				
	Nitzchioid				
	Surirelloid				
Totals					

Field	Type	Count	Total	Taxa Diversity	Biomass
10	Centric				
	Araphid				
	Eunotioid				
	Symmetrical Biraphid				
	Monoraphid				
	Asymmetrical Biraphid				
	Epithemioid				
	Nitzchioid				
	Surirelloid				
Totals					

Field	Richness	Abundance	Species Diversity
1			
2			
3			
4			
5			
6			
7			
8			
9			
10			
Average			
Standard Deviation			

Notes and Questions

Chapter 9

Name: Plankton Field Study

Activity: How to Study Quantitative and Qualitative Plankton Populations.

Level : 3

Developer: Alberto F. Mimo

Site: Outdoors and laboratory

Introduction

Plankton is considered to be any organisms, plant or animal, which is incapable of swimming to and from places on their own and is subject to water currents and wind, water temperatures, solar radiation and other physical actions.

- ✓ There are a number of technical words used in reference to plankton:
- ✓ Metaphyton
- ✓ Phytoplankton
- ✓ Zooplankton
- ✓ Nekton
- ✓ Pleuston
- ✓ Neuston

Use the WEB to look up these words and fit them in to an aquatic environment.

The combined action of phytoplankton and zooplankton on any aquatic system will drive the ecology of the habitat, therefore, extensive studies are conducted to find out how the system will behave at any particular place, season or physical and chemical condition.

Plankton is directly responsible for the primary (Photosynthesis) and secondary (Respiration) productivity of a body of water; therefore scientists are always interested in knowing how much plankton is in the water. This can be done by counting cells, or measuring biomass, or indirectly by measuring chlorophyll *a*. There are also many other methods to obtain numbers in primary and secondary productivity. Scientists have devised extensive methods and techniques and none of these are perfect, but some methods are preferred rather than others.

Finding plankton is easy, plankton is found everywhere, but because it is so diverse, identifying it is the problem. The identification of plankton is a very specialized science, and very specialized taxonomists who are knowledgeable about a variety of groups such as crustaceans, algae, protozoa, and diatoms to name just a few do it. High school students can admire the forms, but to ask them to identify them to the species would be too much to ask and you would never be certain they were right. But plankton can be grouped in phylums or classes and many times even orders and families.

So, what can you do with plankton? One of the easiest tasks is to divide plankton cells into animals or plants: Plankton cells that are responsible for primary or secondary productivity.

In addition, students can also count cells and look at their form, measure them and estimate biomass. It is not something you want to do for fun, but it will provide students with very real numbers they can work with.

Equipment

It all depends on what you will be doing. If you are just interested in collecting plankton, all you need is a plankton net. Very small phytoplankton will slip through the mesh unless it is very fine.

You can drag the net in the water or pour buckets of water through the net. The second choice is the best, as you can measure how much water you have poured through.

Plankton nets are a little expensive but if you want good results you need to buy one. Do not try to make one with a woman's stocking it does not work.

Scientists use a "Van Dorn Water Sample" bottle to collect plankton. They collect one liter and look for plankton by using a filter to discard the water and find the cells. This method is great because it gives you the number of cells per liter directly but it is like looking for a needle in a hay stack. Using a water collecting device you can go to whatever depth you want and test that area. You will need a Sedgewick Rafter glass to count the cells.

You also need microscopes, slides, cover slips, droppers, and dye. If you need to filter the water you also need a filtering system. Millipore® can provide you with a good filtering system.

List to complete qualitative analyses.

Plankton Net

Laboratory glassware (plastic)

Several 200 ml. containers that close (glass)

Markers and labels

Methylene Blue Chloride

Small plastic pipettes

Glass slides with glass cover slips

Stage micrometer .01 mm

Good microscopes to 1000X

Gum arabic

Lugol's solution

List to complete a quantitative analysis

A Van Dorn bottle

Laboratory glassware (plastic)

A Millipore Filtering System

200 ml glass containers that close

Methylene Blue Chloride

Microscope slides with cover slips

Dissecting microscope

Good microscopes to 1000X

Lugol's solution

Sedgewick Rafter Counting Chamber

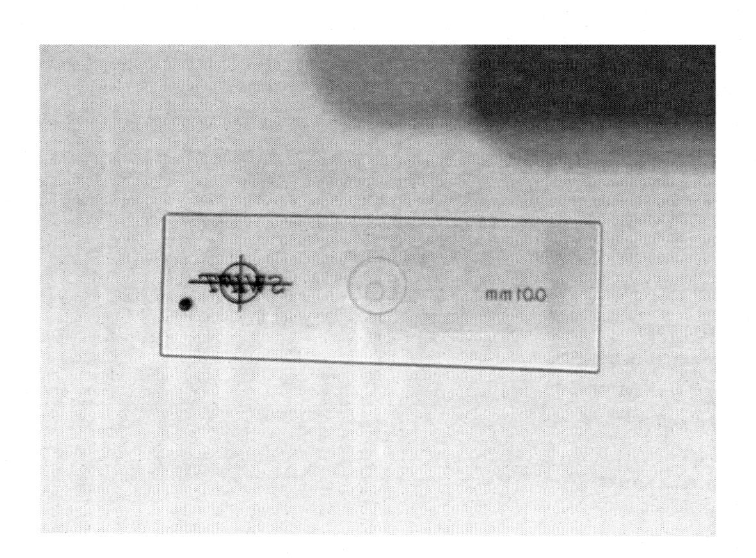

Graduated Micrometer for measuring

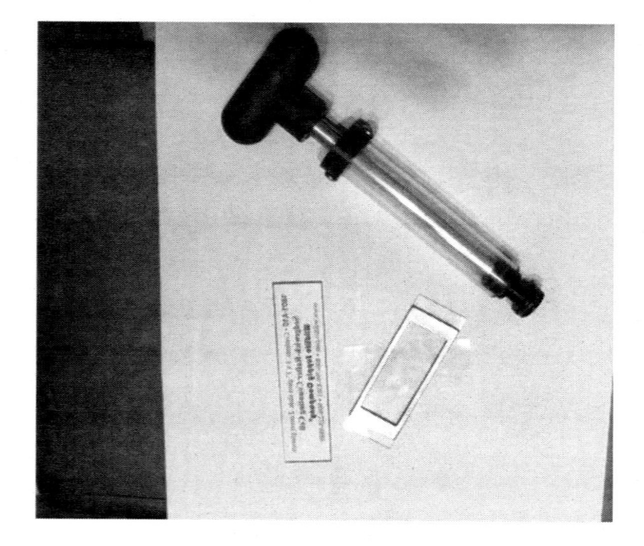

Sedgewick Rafter Counting Chamber and pipette

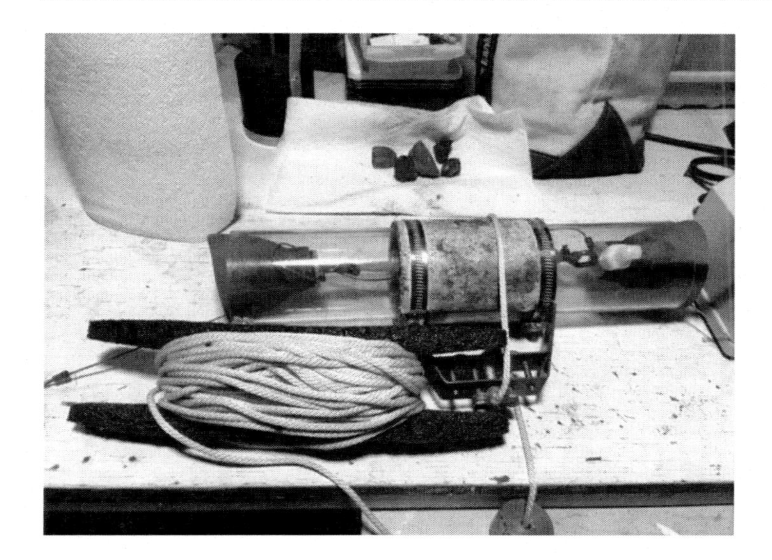

Van Dorn Bottle to collect water

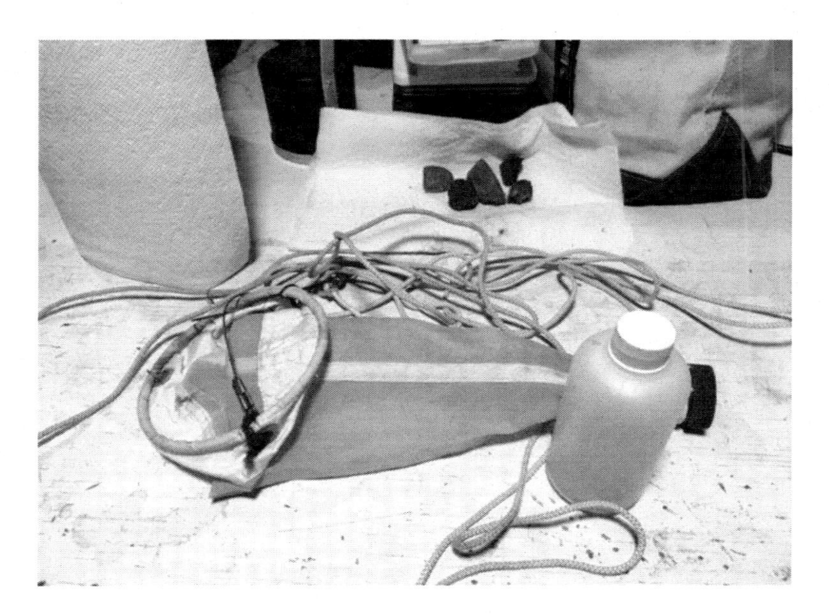

Plankton Net and a plastic container with a top

Identification of Plankton

There are so many good websites that show the plankton and help with their identification that I will only provide you with the sites. This publication is about field techniques, not plant or animal identification.

http://oceandatacenter.ucsc.edu/home/outreach/PhytoID_tabletformat.pdf

https://sealevel.jpl.nasa.gov/files/archive/activities/ts3ssac3.pdf

https://www.planktonportal.org/

https://www.marine.usf.edu/pjocean/packets/f97/plank_id.pdf

https://westerndiatoms.colorado.edu/

http://cfb.unh.edu/cfbkey/html/

Methods for Qualitative Analyses

Best locations to get your collection are ponds and sea shores. A scientist uses a boat to tow the plankton net back and forth at a particular depth to obtain the sample. You probably do not have a boat, so we use the bucket brigade method or the throw and retrieve method.

Bucket brigade method – Get your students in a row with most inside the water and give them 5 buckets. The last in line, the one that is at a deeper location collects the water and passes the bucket to the next in line. They pass the bucket all the way to the end and the last student empties the bucket inside the plankton net. You need about 25 gallons to get a good sample. Do this several times and keep the samples separate. Do not empty the bucket next to the place where you collect the water.

Throw and retrieve method. Tie a long rope to the net and throw the net in the water as far as you can, slowly retrieve the net, and throw it again. Do this between 25 and 50 times. Take several samples.

Once you have several good plankton samples it is time to fix them with Lugol's solution. Lugol's solution can be purchased or you can make your own batch. It is a solution of potassium iodide with iodine in water.

Lugol's is available in various strengths from 1% to slightly less than 13% iodine (wt/v). The most commonly used 15% solution consists of 5% (wt/v) elemental iodine (I_2) and 10% (wt/v) potassium iodide (KI) mixed in distilled water, and has a total iodine content of 142.5 mg/mL (126.5 mg/g). The iodide combines with elemental iodine to form a high concentration of potassium triiodide (KI_3) solution. Based on Instructions from Wikipedia.

Lugol's is good because it stains the plankton and makes it sink.

You can also buy Lugol's through the web.

Microscope method

Shake the sample and place one drop on a slide. Cover it with a slip. Use glass, it works much better than plastic. Look at the slide, choose one organism which is dead and draw it. It is not a bad idea to bring a live sample from the field and look at the organism alive!.

To begin the process ask the students to concentrate on one organism. Ask them to use the web, on a computer or their phones, to identify it. Then it is a matter of getting to know what it is, its natural history and all they can find out about this microscopic thing. Once the students are ready, get them to present a brief report to the whole class. Start to complete a picture of what plankton is and its diversity. Students can take pictures using their cell phones.

Making permanent slides

Not all the planktonic organisms do well. Some get destroyed due to osmosis and to other physical pressures, but here is how you make a permanent slide.

1. Find the organism you want to preserve.
2. Using a dissecting scope lift the organisms with a pipette and place it on another slide. There will be some water.
3. You may get more than one thing.
4. Using a pipette get a drop of gum arabic, maybe half a drop, and place it in the center of the slide where the plankton is.
5. Cover the slide with a cover slip. Press a little to spread the gum everywhere.
6. Using a clear nail polish, cover the border of the slide.
7. Let the whole thing dry.

Taking measurements

To measure the organisms you can use a stage micrometer. This is basically a very small ruler that you can place on the microscope and measure the size of the organism. Micrometers cost from $10 to $100. Do not spend too much.

Plankton Biomass

Productivity of a lake is a major issue. More plankton growth means that the lake is more productive. This especially refers to Phytoplankton, but Zooplankton is also important.

Measuring biomass is a way of measuring "Carbon" productivity. Each invividual phytoplankton cell is made out of carbon atoms. So, in order to measure biomass you need to measure the cells' volume of each individual organism. Not an easy task but possible.

- The first task would be to collect a known volumetric sample, say a liter. Once you have that liter, the amount of plankton inside is small, so this liter needs to be filtered or, using a centrifuge, the solids need to be separated.
- Using a Millipore filters of 47um and a Millipore filtering system is the best way to complete the separation.
- Reduce the initial sample by 90%.
- Now use a Sedgewick Rafter Glass and pipette to look at the individual cells.
- Identify the different species of plankton and count the number of each species found.
- The next task is to isolate each species and look at the shape of the cells individually. Compare them to known geometric forms.
- Using a graduated micrometer calculate the volume of each different cell. You will need to transfer the plankton to a regular microscope slide.
- Use the formulas shown below.
- Multiply the volume of one cell times the number of cells present with similar shapes. See the form below.
- Use the data sheet to count and measure plankton. Formulas are shown separately.

Here are some examples

Species	Shape	Max length	Volume	Surface area
Cerratium	circular	201 um.	43740 um3 30000 cells	9600 um^2
Anabeana	filamentous	60 um. 20 cells	2040 um3	2110 um2
Tabellaria	global	96 um. 8 cells	13800 um^3 10.000 calls	9800 um^2
Volvox	spherical	450	$47 \times 10^6 um^3$	$636 \times 10^3 um^2$

Shape of Organisms Number	Taxa, Family, Genus or Sp.	Number of Cells	A: B: C: D: r:	Volume	Number of Organisms	Total Biomass

- Research Team: ______________________________
- Sample #: Location: Town:
- Collection Method:
- Depth:
- Date:
- Time:

•

Formulas

Shape	Volume	
Sphere	$V = 4/3\ \pi r^3$	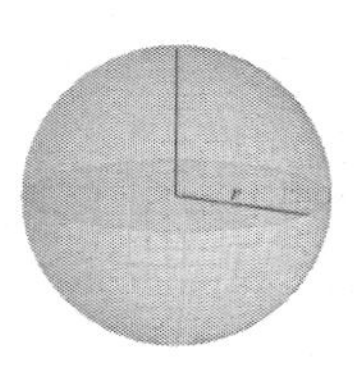
Ellipsoid	$V = 4/3\pi abc$	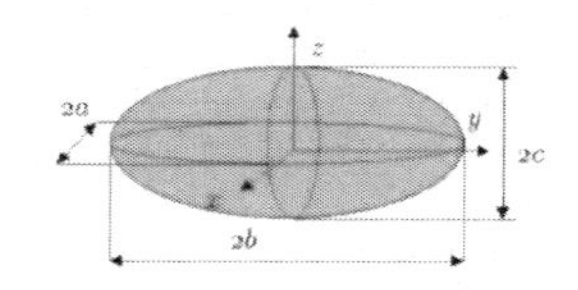
Cylinder	$V = \pi r^2 h$	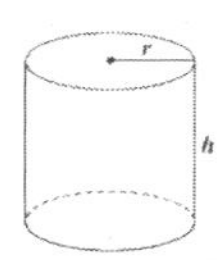
Cylinder + 2 Spheres	$V = \pi b^2\ (a/4-b/12)$	
Cylinder + 2 Cones	$V = (\pi/)(b^2)(a-b/3)$	
Cone	$V = (\pi r^2)(h/3)$	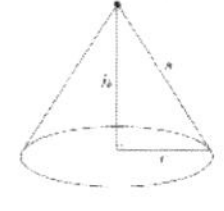
Double Cone	$V = (\pi/12)(a)(b^2)$	
Rectangular Box	$V = (a)(b)(c)$	

Getting Quantitative Data

This data is much more difficult to get mainly because it requires more equipment and more steps to complete the mission.

Samples should be taken from 3 areas within the lake, one on the north, one in the center and one on the southern end of the lake. All samples are collected at deep areas and the complete location is recorded by use of a GPS.

Using a Van Dorn bottle a one-liter sample of water is taken at 1, 2, 3, 4, and 5 meters. All samples are placed in a bucket and mixed for 5 minutes with a large wooden stick. After the sample has been mixed, three, one liter samples are placed in collecting bottles and fixed using Lugol's solution.

At the same site, a one-foot in diameter standard plankton net will be lowered to a depth of 5 meters and brought back to the surface. All plankton collected should be collected in a small vial for plankton identification. Plankton should be fixed using Lugol's solution to prevent predation and denaturalization.

Dealing With the Sample

The sample taken with the plankton net will be studied using the method mentioned before. The one liter samples need to be filtered using a Millipore filtering system.

Place the sample in the receiving cup and, using the syringe, suck up the extra water. The filter used should be a 47um filter.

Take enough water out to leave only about 100ml. This is your final sample.

Microscope Work

Using the Sedgewick pipette take a .5 ml sample and place it on the slide, which has a pool that fits exactly .5 ml.

Using the lowest magnification on the microscope, count the number of cells you see on the slide going back and forth to cover the whole .5 ml.

See laboratory sheet.

A = Volume of water filtered = 1000 ml

B = Volume of water collected = 1000 ml

C = ml subsample = 100 ml

D = Plankton count subsample

E = Plankton count sampled

E/A = Plankton concentration on the lake

Lab Data Sheet

Date:	School:

PLANKTON COUNT & CONCENTRATION DATA

Sample	Plankton Count Subsample D	mL's Subsampled C	Plankton Count Sample D*B/C = E	Plankton Concentration E /A
Plankton 1				
Plankton 2				
Plankton 3				
Grand Total				

Sample # **1 2 3**

A: Vol. of Water Filtered = **100L**/sample **B**: Vol. of Water Collected (**mL**):

Alternative method
If you want a simpler method, you can filter a number of liters of water through the plankton net and use the alternate method where:
A = Water filtered = for example 25 liters
B = Water collected in a small plastic container 200 ml
C = Total subsample 1 ml
To count the plankton place on slides with a 1 ml pipette until you use the whole 1 ml sample.
D = Total Plankton Count in all the drops.

Date:	School Name:	Sample: 1 or 2 or 3 (**circle one**)

Record Number of Plankton cells seen in each drop of water for each mL **subsampled**

Drop #	mL 1 PHYTO	mL 1 ZOO	mL 2 PHYTO	mL 2 ZOO
1				
2				
3				
4				
5				
6				
7				
8				
9				
10				
11				
12				
13				
14				
15				
16				
17				
18				
19				
20				
21				
22				
23				
24				
25				
26				
27				
28				
29				
30				
31				
32				
33				
34				
35				
TOTAL				

Plankton Concentration Calculation

Total Plankton Count in Sample = **E** = **(D * B) / C**

E/A

<u>Total Plankton * Voumel of Sample (mL)</u>

Volume of Subsample (mL)

Plankton Concentration = E / A

<u>Total Plankton Count in Sample</u>

Total Volume of Water Filtered (L)

A – Water Filtered		**B** – Water Collected	
C – Total # ml		**D** -- Total Plankton	

See Field Data Sheet for Values "**A**" and "**B**" Above

Plankton Identification Data Sheet

Date:	School Name:	Sample: 1 or 2 or 3
	(circle one)	

Plankton Identification

(Draw & Name Each Organism Seen in the Boxes Below)

Phytoplankton (PLANTS)	*Zooplankton (ANIMALS)*

Chapter 10

Name: Field and Laboratory Methods in Micropaleontology

Activity: Study of microfossil samples collected from oceans, bogs and ponds that were deposited in the past.

Level: 3

Developer: Alberto F. Mimo

Site: Outdoors and Laboratory

Introduction

Scientists have been able to collect large amounts of information about the past by looking at microfossil deposits found deep in the oceans, on the shore, in lakes and ponds, and in bogs.

As time goes by, pollen and spores from plants and small microorganisms living in water are deposited on the surface of sediments. These organic pieces become part of in the sediments and accumulate with time; organic matter not only accumulates but changes as the plant communities and the organisms in water change.

With time and patience, one can take a sediment core from the shore, lake, pond or bog and look for clues.

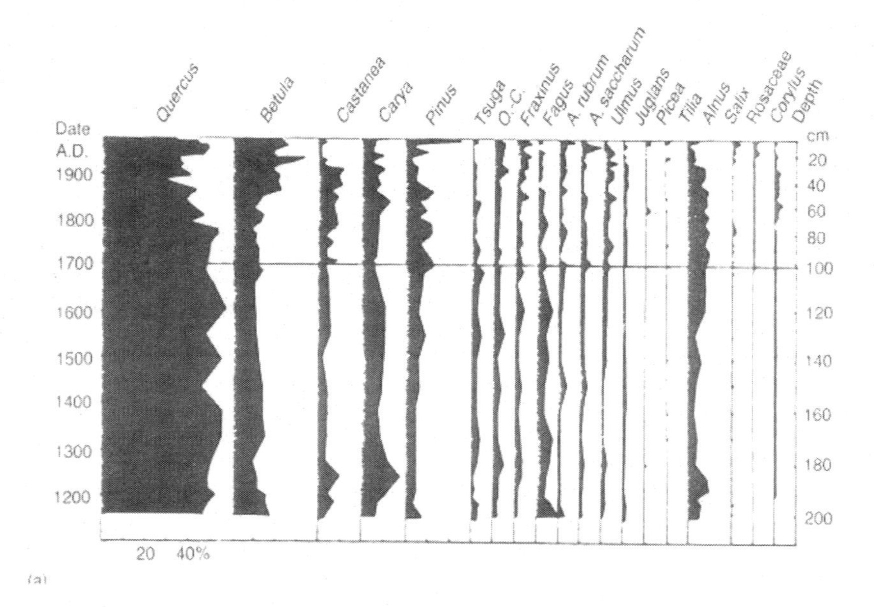

Careful observations and counts of pollen and spore grains will provide us with a picture of the different plant communities that were found throughout the years.

To complete a careful study of a core sample is beyond the scope of this Field Techniques manual but we can explore the techniques that are available and have a better understanding of what it takes to complete one of these studies.

- ✓ Soil samples contain a variety of organisms:
- ✓ Foraminifera (protozoans in the class rhizopods)
- ✓ Ostracods (Subphylum Crustacea)
- ✓ Pteropods (Class Gastropoda)
- ✓ Palynomorphs (pollen, seeds)

- ✓ Dinocysts (Dinoflagellates)
- ✓ Calcareous nannofossils (coccoliths and coccospheres of haptophyte algae)
- ✓ Diatoms (Bacillariophyta, cell wall composed of hydrated silica)

Each of these organisms requires different methods for cleanup and counting. Some of these cleanup methods use very harsh and dangerous chemicals. Those methods will not be covered here (such as methods using hydrofluoric acid and carbon tetrachloride). The majority of these organisms do not require much preparation and they can be seen with the microscope, so they will be covered here.

Field Sampling Methods

Location

One can obtain a core sample from just about any body of water, but it will be more interesting to compare two samples and see if there are any differences.

Select two ponds that are nearby and within your community or, even better, select one pond and a bog. It may be necessary for you to travel far to collect the sample from the bog but this is something you can do (without your students) during the weekend. Bogs are really good for this type of sampling.

Sample

You will need to obtain an Auger to get good samples at reasonable depths. Most scientists will take a two or three-meter sample. For your sampling, I would recommend taking only 25 cm. which will be enough to do most of the work.

Plastic tubes in different sizes and rubber tops

Using a plastic tube that you can close on the top to make a vacuum is best. Hammer the plastic tube into the sediment, close the top using a rubber top and, when you are ready, pull the core sample out. Because the sample is under water it has a tendency to crumble. You will need to place the sample on a tray to keep it from crumbling, and reconstruct its shape as best as you can.

Soil sample next to Munsell color chart

As you take the sample it will also be a good idea to take some photos of the core and of the site. If you have a GPS take a reading so that you can record latitude and longitude. These days any cell phone has a GPS.

Processing the Sample

Once you have the sample you need to start completing the sample form included in this manual. Some of that information you can complete in the field, and some of the information will be completed as you analyze the sample.

Samples collected on a saltmarsh

Sample Preparation for Foraminifera, Ostracods, and Pteropods

Microfossils are commonly used by paleontologists and marine biologists. Foraminifera is an exclusive marine fossil and good indicator of periods from the Jurassic to the present. Benthic foraminifera is an indicator of Cambrian species, and stains with Rose Bengal.

Ostracods are crustaceans, they have a shell and they are very small, you can see the legs sticking out on the shell edge. These organisms are found in marine, lacustrine and terrestrial habitats and occur from the Cambrian era to the present.

Pteropods have an aragonitic shell (aragonite is a crystalized form of calcium carbonate) ranging from 10^2 to 10^4 Um in size. They are very small. These are found exclusively in marine environments and occupy mesopelagic zones of the ocean.

Method

1. Chemicals will destroy the carbonate shells on these organisms, so the preparation is simple. Sieve the sample in a #230 sieve, 63 um, to separate the sample from silt and clay. The part of the sample that goes through the sieve is silt and clay, the catch is what you want.
2. Just rinse the sample using a funnel and filter paper using distilled water. Do this at least three times.
3. Dry using a blow dryer. Use a small brush to sort the samples, and then use a stereo microscope to look for the organisms.

Diatoms

Diatoms are covered with silica and have two shells called frustules that will break open just like a petri dish when it falls on a table. Diatoms need to be cleaned, and to collect them from the sample they need to be centrifuged.

Cleaning Diatoms

Here is a way to clean the diatoms that is not dangerous for your students. After the field trip, take a portion of your sample (about 40 %) and clean the diatoms.

Method

1. To clean the diatoms you will use a centrifuge to separate the diatoms from the liquid. Shake the sample well and place some sample in the centrifuge tube. You need to fill 4 tubes as the centrifuge needs to be balanced.
2. Spin the samples for 4 minutes. At the end you will be able to see the diatoms at the bottom of the tube.
3. Using a plastic pipette, take the water out. Leave the diatoms.
4. Add Clorox© to the sample, about half way up the tube. Mix it with the diatoms and let the tube rest for 20 minutes.
5. Spin the sample, and take out the Clorox. Leave the diatoms in and add more Clorox. Let the tube rest another 20 minutes. It is like doing laundry.
6. Take the Clorox© out and then clean the diatoms by using distilled water. Add distilled water, mix, and spin.
7. Repeat the process 3 times.
8. Place in a clean container all the newly cleaned diatoms with some distilled water. This is your clean sample.

Identification

As you can see, there are different organisms that have been processed in different ways, so you need to investigate each separate group in a different way.

1. Place a drop of clean diatoms on a slide.
2. Look for a diatom
3. Let your students draw it and identify the diatom to the genus using a diatom key. You should have no problem seeing these diatoms clearly, with all their details.

We would like to get a list of the first 100 diatoms you see with drawings and your best ID. That means each student will do 3 diatoms. Diatoms should be selected at random. Use http://westerndiatoms.colorado.edu as a resource.

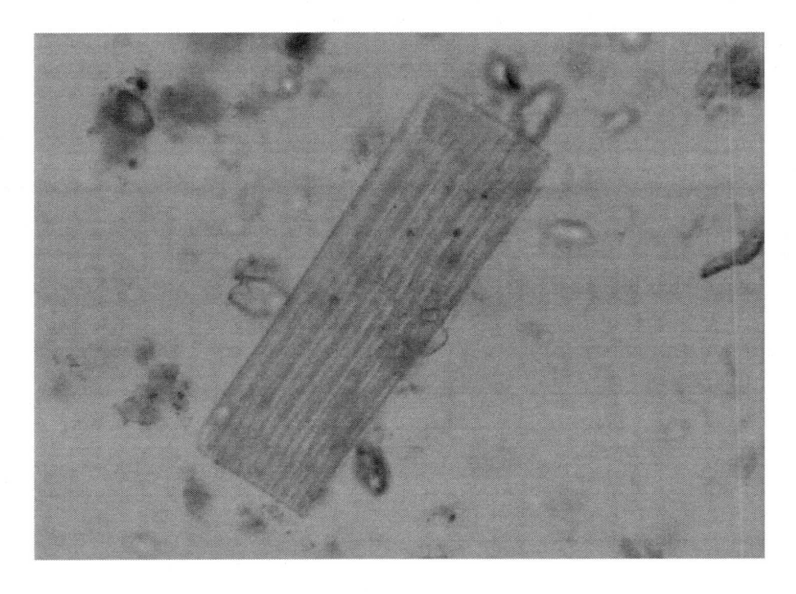

This sample contains some diatoms.

Foraminifera and Other Organisms Containing Calcium Carbonate

Below we provide you with a short set of examples of microfossils you may find in your sample. Remember that the objective of this field technique is to enumerate types and abundance of different microfossils found at different depths in your core sample. It will get your students acquainted with field and laboratory techniques and help them to understand that these are an extensive number of organisms found on the earth.

These websites will help you with that task.

http://www.foraminifera.eu/taxo.php

http://westerndiatoms.colorado.edu/

http://www.discoverlife.org/mp/20q?guide=Pollen

http://www.microlabgallery.com/PollenFile.aspx

Color

You can analyze the color of each sample by using the MUNSELL color chart or by using the color chart here. (This color chart was taken from the Project Oceanology CD Investigating the Marine Environment).

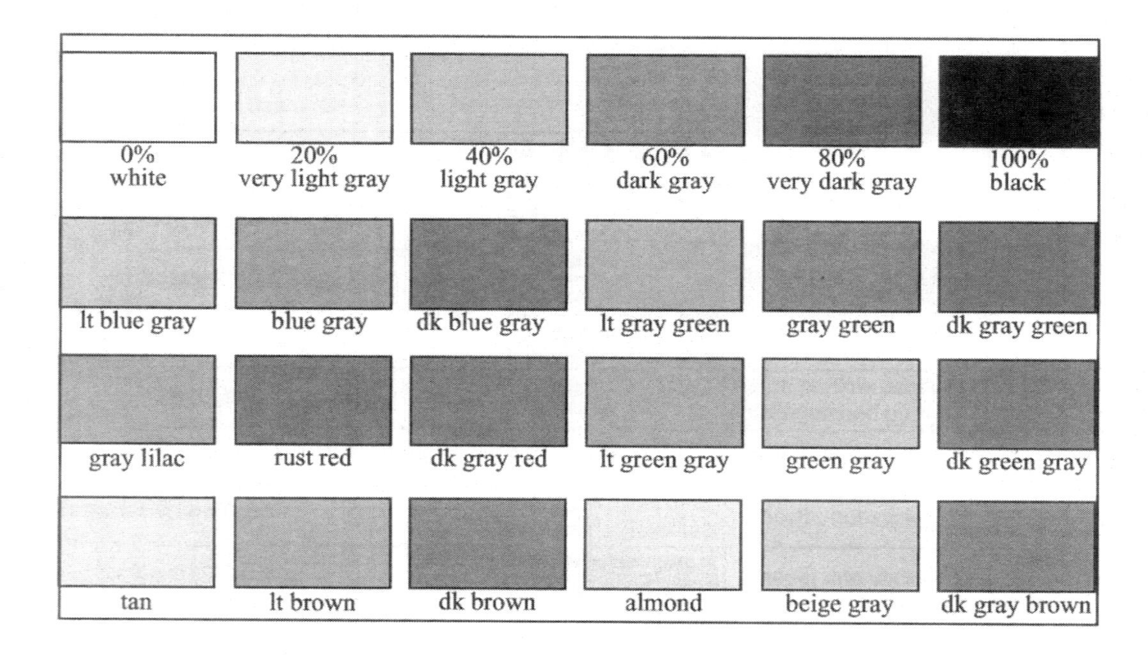

Core Sample Data Sheet

Date:____________________

Name of Location: _____________________________

Town: _________________ County: _______________________

Exact Location:____

Lat: ______________ Long: ____________________

Transect #: _________ Quadrat #: __________

Sample #: _______ Core Total Length: _________

Sample Number: ____________________ Weight: __________________

Color: _____________________ Odor: ______________________

% of Organic/ Sand/ Silt/ Clay: _____ _____ _____ _____

Identify what groups were represented in your sample and the number of cells you saw.

Group	Slide #1	Slide # 2	Slide # 3	Slide #4	Slide #5	Slide #6	Slide #7	Slide # 8	Slide #9	Slide # 10	Total
Foraminifera											
Ostracoda											
Radiolaria											
Diatoms											
Conodonts											
Calpionellids											
Coccoliths Nannoliths Chitinozoa											
Prasinophyceae Algae											
Acritarchs											
Dinoflagellate Cysts											
Spores											
Pollen											
Arthropods											
Other											

Microfossils

Below, we provide you with a short number of examples of microfossils you may find in your sample. Remember that the objective of this field technique is to enumerate types and abundance of different microfossils found at different depths in your core sample.

These websites will help you with that task.

http://www.geotop.ca/upload/files/laboratoires/laboratoire-de-micropaleontologie-et-palynologie-marine-uqam/Micropal_Methods_2010.pdf

http://www.foraminifera.eu/taxo.php

http://westerndiatoms.colorado.edu/

http://www.discoverlife.org/mp/20q?guide=Pollen

http://www.microlabgallery.com/PollenFile.aspx

Identification Task: Using your computer, identify on the web the following Microfossils found in your samples. Please take a good look at the pictures. They will be of great help in your identifications.

Foraminifera

____

____

____

____

Ostracods

____

____

____

____

Diatoms

Radiolarians

Conodonts

Calpionellids

Coccoliths

Nannoliths

Dinoflagellate Cysts

Acritarchs

Prasinophycean algae

Freshwater algae

Chitinozoan

Spores

Pollen

Arthropods

BIBLIOGRAPHY

Eds. Baker J. M., & W.J. Wolf. 1987. Biological Surveys of Estuaries and Coasts. Cambridge University Press. ISBN 0521 311918

C. E. Epifanio*, A. I. Dittel, S. Park, S. Schwalm, A. Fouts. 1998 Early life history of *Hemigrapsus sanguineus, a* non-indigenous crab in the Middle Atlantic Bight (USA). Ecology Progress Series: Volume 170: 231-238

Cox G. W., Laboratory Manual of General Ecology. 1967. Wm. C. Brown Company Publishers. ISBN 0-697-04679-6

Habraken J., Microsoft Office 2010 in Depth. 2011, Que Publishing. ISBN 978-0-7897-4309-1

Jensen G. C., P. S. McDonald, & David A. Armstrong. 2002. East meet west :Competitive interactions between green crab Carcinus maenas and native and introduces shore crab Hemigrapsus spp. Marine Ecological Progress Series: Vol. 225: 251-262.

Lack D.1968. Population Studies of Birds. Oxford University Press. ISBN 13: 9780198573357

Ledesma M. E. & Nancy J. O'Connor. 2003. Habitat and Diet of the Non-Native Crab Hemigrapsus sanguineous in Southeastern New England. Northeast Naturalist: Vol 8, No. 1, pp 63-78.

Ludwig J. A., & James F. Reynolds. 1988. Statistical Ecology. Wiley International Publications. John Wiley & Son .ISBN 0-471-83235-9

Magurran A. E. 1988. Ecological Diversity and Its Measurements. Princeton University Press. ISBN 0-691-08491-2

McDermott. J. 1991. A Breeding Population of the Western Pacific Crab Hemigrapsus sanguineous (Crustacea: Decapoda: Grapsidae) Established on the Atlantic Coast of North America. Biol. Bull. 181: 195-198.

Mimo A. F. 2002. Fresh Water Crustaceans of Connecticut. The Amateur Naturalist ISBN 0-9741411-0-0

Needham J. G.,& Paul Needham. 1962. A Guide to the Study of Freshwater Biology. Holden-Day, Inc. ISBN 0-8162-6310-8

Reynolds C. S.1984. The Ecology of Freshwater Phytoplankton. Cambridge University Press. ISBN 0-521-28222-5

Scholander, Hock, Waters, Johnson and Irving 1950: Biol. Bull., 99:254

Stephens L. J. 2006. Beginning Statistics. McGraw-Hill Companies Inc. ISBN 0-07-145932-4

Walker C. Leslie, & Charles E. Roth. 2000, Keeping a Nature Journal., Storey Publications, ISBN 978-1-58017-493-0

Wetzel R. G., & Gene E. Likens. 1990. Limnological Analyses. Spring-Verlang. ISBN 0-387-97331-1.

Wood P. 1982. Scientific Illustrations. Van Nostrand Reinhold Company. ISBN 0-442-29307-0

Websites

https://www.homesciencetools.com/content/reference/IN-INSEPIN.pdf

https://www.youtube.com/watch?v=MT5VGISCtg4

http://extension.oregonstate.edu/umatilla/sites/default/files/PINNING__INSECTS.pdf

https://bugguide.net/node/view/36900

https://www.youtube.com/watch?v=upJ03FCE2RI

http://www.organiclawndiy.com/2009/06/make-your-own-soil-sampler.html

https://westerndiatoms.colorado.edu/.

http://oceandatacenter.ucsc.edu/home/outreach/PhytoID_tabletformat.pdf

https://sealevel.jpl.nasa.gov/files/archive/activities/ts3ssac3.pdf

https://www.planktonportal.org/

https://www.marine.usf.edu/pjocean/packets/f97/plank_id.pdf

https://westerndiatoms.colorado.edu/

http://cfb.unh.edu/cfbkey/html/

http://www.foraminifera.eu/taxo.php

http://westerndiatoms.colorado.edu/

http://www.discoverlife.org/mp/20q?guide=Pollen

http://www.microlabgallery.com/PollenFile.aspx

Alberto Mimo graduated with a Master's Degree in biological sciences from Central Connecticut State University, and has been living in Connecticut ever since. He has spent the last 35 years working in the field of Environmental Education for the Connecticut Department of Environmental Protection. He has designed and coordinates five major statewide environmental education programs, mostly to provide technical education to high school students and adults. One of his programs "SEARCH", a water quality monitoring education program, was chosen and funded by the National Science Foundation to provide systemic change for biology and chemistry in all the high schools in Connecticut. Alberto has been recognized as providing Connecticut with a number of very unique environmental education programs with applications beyond his home state. He is the Recipient of the Connecticut Outdoor and Environmental Education Association "Environmental Educator of the Year" award in 1989, the 1991 National "Roger Tory Peterson" environmental education award and the 1995 Environment 2000 Governor's award. In 1995 he was nominated by his peers and also awarded the DEP Distinguished Service Award based on his outstanding contributions to the Department of Environmental Protection. The Connecticut Audubon Society also awarded him the 1997 "Piping Plover" appreciation award for his contribution to the Audubon Society in Connecticut. In 1999, Alberto was honored once more by Briarwood College with their "First Environmental Service Teachers award". In 2011 he was awarded the Dr. Sigmund Abeles Science Award for his outstanding service to Science Education in the State of Connecticut and in 2015 he was inducted into the Science National Honor Society.

Alberto believes that the best way to influence students and reach their hearts is by providing them with an opportunity to do research and monitor the environment.

Anna M. Jalowska research interests are directed at the effect of anthropogenic global change on the hydrologic cycle, and fluvial and coastal environments. She holds a M.S. in Hydrology from University of Adam Mickiewicz in Poznan, Poland and a Ph.D. in Marine Sciences from University of North Carolina in Chapel Hill, NC in the US. Her current and past research allowed her to gain expertise in effects of climate change, land use change, river damming and sea-level rise on coastal and riverine environments.

Jennifer Aley has edited books, reports, and many articles on social and legal environmental subjects. She has fifteen years of experience as a park naturalist and environmental educator. She designed experience based science and natural history programs for school groups and the general public, and edited and wrote articles for two newsletters that provided educational information on environmental subjects. She has an MES from the Yale School of Forestry and Environmental Studies, an MA in Sociology from George Mason University and a BA in Geography from Central Connecticut State University. She enjoys painting, hiking and birdwatching.